NIST Handbook 133
Checking the Net Contents
of Packaged Goods

as adopted by the 95th National Conference on Weights and Measures 2010

Editors:

Linda Crown
David Sefcik
Lisa Warfield

Carol Hockert, Chief
NIST Weights and Measures Division
Gaithersburg, MD 20899-2600

U.S. Department of Commerce
Gary Locke, Secretary

**National Institute of
Standards and Technology**
Patrick D. Gallagher, Director

NIST Handbook **133**

January 2011

Supersedes
Fourth Edition, January 2005

National Institute of Standards and Technology Handbook 133, 2011 Edition Natl. Inst. Stand. Technol. Handb. 133, 2011 Ed. 188 Pages (Jan. 2011)

CODEN : NIHAE2

U.S. GOVERNMENT PRINTING OFFICE
WASHINGTON: 2011

Internet: bookstore gpo gov Phone: toll free (866) 512-1800; DC area (202) 512-1800 Fax: (202) 512-2250 Mail: Stop SSOP, Washington, DC 20402-0001 **ISBN** 0-16-051249-2

Foreword

This handbook has been prepared as a procedural guide for the compliance testing of net content statements on packaged goods. Compliance testing of packaged goods is the determination of the conformance of the results of the packaging, distribution, and retailing process (the packages) to specific legal requirements for net content declarations. This handbook has been developed primarily for the use of government officials. However, it should also be useful to commercial and industrial establishments in the areas of packaging, distribution, and sale of commodities.

In conducting compliance testing, the conversion of quantity values from one measurement system to another (e.g., from the metric system to the avoirdupois system) should be handled with careful regard to the implied correspondence between the accuracy of the data and the number of digits displayed. In all conversions, the number of significant digits retained should ensure that accuracy is neither sacrificed nor exaggerated. For this 2010 edition of Handbook 133 all dimensions for test procedures, devices, or environments have been rounded to two significant digits (e.g., 2.5 cm to 1.0 in) or to a precision level applicable to the test equipment (e.g., 200 kPa for 25 psi and 35 MPa for 5,000 psi).

THIS PAGE INTENTIONALLY LEFT BLANK

Table of Acronyms

Acronym	Term
AMAV	Adjusted Maximum Allowable Variation
ANGW	Adjusted Nominal Gross Weight
AOSA	Association of Official Seed Analysts
ASEL	Adjusted Sample Error Limit
ASTM	American Society for Testing Materials International
CFR	Code of Federal Regulations
CGA	Compressed Gas Association
EPA	Environmental Protection Agency
FDA	Food and Drug Administration
FDCA	Food, Drug and Cosmetic Act
FPLA	Fair Packaging and Labeling Act
FSIS	Food Safety and Inspection
FTC	Federal Trade Commission
HB 130	NIST Handbook 130 " *Uniform Laws and Regulations in the areas of Legal Metrology and Engine Fuel Quality"*
HB 133	NIST Handbook 133 *"Checking the Net Contents of Packaged Goods"*
HB 44	NIST Handbook 44 *"Specifications, Tolerances, and Other Technical Requirements for Weighing and Measuring Devices."*
LNQC	Labeled Net Quantity of Content
MA	Moisture Allowance
MAV	Maximum Allowable Variation
MSDS	Material Safety Data Sheets
NGW	Nominal Gross Weight
NIST	National Institute of Standards and Technology
UME	Unreasonable Minus Errors
SEL	Sample Error Limit
TTB	Alcohol and Tobacco Tax and Trade Bureau
UPLR	Uniform Packaging and Labeling Regulation
USDA	U.S. Department of Agriculture

THIS PAGE INTENTIONALLY LEFT BLANK

Table of Content

Chapter 1. General Information

1.1. Scope

Routine verification of the net contents of packages is an important part of any weights and measures program to facilitate value comparison and fair competition. Consumers have the right to expect packages to bear accurate net content information. Those manufacturers whose products are sold in such packages have the right to expect that their competitors will be required to adhere to the same laws and regulations.

The procedures in this handbook are recommended for use to verify the net quantity of contents of packages kept, offered, or exposed for sale, or sold by weight, measure (including volume, and dimensions), or count at any location (e.g., at the point-of-pack, in storage warehouses, retail stores, and wholesale outlets).

a. When and where to use package checking procedures?

An effective program will typically include testing at each of the following levels.

(1) Point-of-pack

Testing packages at the "point-of-pack" has an immediate impact on the packaging process. Usually, a large number of packages of a single product are available for testing at one place. This allows the inspector to verify that the packer is following current good packaging practices. Inspection at the point-of-pack also provides the opportunity to educate the packer about the legal requirements that products must meet and may permit resolution of any net content issues or other problems that arise during the testing. Point-of-pack testing is not always possible because packing locations can be in other states or countries. Work with other state, county, and city jurisdictions to encourage point-of-pack inspection on products manufactured in their geographic jurisdictions. Point-of-pack inspections cannot entirely replace testing at wholesale or retail outlets, because this type of inspection does not include imported products or the possible effects of product distribution and moisture loss. Point-of-pack inspections only examine the manufacturing process. Therefore, an effective testing program will also include testing at wholesale and retail outlets.

(2) Wholesale

Testing packages at a distribution warehouse is an alternative to testing at the point-of-pack with respect to being able to test large quantities of and a variety of products. Wholesale testing is a very good way to monitor products imported from other countries and to follow up on products suspected of being underfilled or underweight based on consumer complaints or findings made during other inspections, including those done at retail outlets.

(3) Retail

Testing packages at retail outlets evaluates the soundness of the manufacturing, distributing, and retailing processes of the widest variety of goods at a single location. It is acceptable and practical for weights and measures jurisdictions to monitor packaging procedures and to detect present or potential problems. Generally, retail package testing is not conducive to checking large quantities of individual products of any single production lot. Therefore, follow-up inspections of a particular brand or lot code number at a number of retail and wholesale outlets, and ultimately at the point-of-pack are extremely important aspects in any package-checking scheme. After the evaluation of an inspection lot is completed, the jurisdiction should consider what, if any, further investigation or follow-up is warranted. At the point-of-sale, a large number of processes may affect the quality or quantity of the product. Therefore, there may be many reasons for any inspection lot being out of compliance. A shortage in weight or measure may result from mishandling the product in the store, or the retailer's failure to rotate stock. Shortages may also be caused through mishandling by a distributor, or failure of some part of the packaging process. Shortages may also be caused by moisture loss (desiccation) if the product is packaged in permeable media. Therefore, being able to determine the cause of an error in order to correct defects is more difficult when retail testing is used.

(Amended 2010)

b. What products can be tested?

Any commodity sold by weight, measure, or count may be tested. The product to be tested may be chosen in several ways. The decision may be based on different factors, such as (1) marketplace surveys (e.g., jurisdiction-wide surveys of all soft drinks or breads), (2) surveys based on sales volume, or (3) audit testing (see Section 1.3. "Sampling Plans") to cover as large a product variety as possible at food, farm, drug, hardware stores, or specialty outlets, discount and department stores. Follow-up of possible problems detected in audit testing or in review of past performance tends to concentrate inspection resources on particular commodity types, brand names, retail or wholesale locations, or even particular neighborhoods. The expected benefits for the public must be balanced against the cost of testing. Expensive products should be tested because of their cost per unit. However, inexpensive items should also be tested because the overall cost to individual purchasers may be considerable over an extended period. Store packaged items, which are usually perishable and not subject to other official monitoring, should be routinely tested because they are offered for sale where they are packed. Products on sale and special products produced for local consumption should not be overlooked because these items sell quickly in large amounts.

Regardless of where the test occurs, remember that it is the inspector's presence in the marketplace through routine unannounced testing that ensures equity and fair competition in the manufacturing and distribution process. Finally, always follow-up on testing to ensure that the problems are corrected; otherwise, the initial testing may be ineffective.

1.2. Package Requirements

The net quantity of content statement must be "accurate," but reasonable variations are permitted. Variations in package contents may be a result of deviations in filling. The limits for acceptable variations are based on current good manufacturing practices in the weighing, measuring, and packaging process. The first requirement is that accuracy is applied to the average net contents of the packages in the lot. The second requirement is applied to negative errors in individual packages. These requirements apply simultaneously to the inspection of all lots of packages except as specified in Section 1.2.(6) "Exceptions to the Average and Individual Package Requirements."

(1) Inspection Lot

An "inspection lot" (called a "lot" in this handbook) is defined as a collection of identically labeled (except for quantity or identity in the case of random packages) packages available for inspection at one time. The collection of packages will pass or fail as a whole based on the results of tests on a sample drawn from the lot. This handbook describes procedures to determine if the packages in an "inspection lot" contain the declared net quantity of contents and if the individual packages' variations are within acceptable limits.

(2) Average Requirement

In general, the average net quantity of contents of packages in a lot must at least equal the net quantity of contents declared on the label. Plus or minus variations from the declared net weight, measure, or count are permitted when they are caused by unavoidable variations in weighing, measuring, or counting the contents of individual packages that occur in current good manufacturing practice. Such variations must not be permitted to the extent that the average of the quantities in the packages of a particular commodity or a lot of the commodity that is kept, offered, exposed for sale, or sold, is below the stated quantity. (See Section 3.7. "Pressed and Blown Glass Tumblers and Stemware" and Section 4.3. "Packages Labeled with 50 Items or Fewer" for exceptions to this requirement.)

(3) Individual Package Requirement

The variation of individual package contents from the labeled quantity must not be "unreasonably large." In this handbook, packages that are underfilled by more than the Maximum Allowable Variation specified for the package are considered unreasonable errors. Unreasonable shortages are not generally permitted, even when overages in other packages in the same lot, shipment or delivery compensate for such shortage. This handbook does not specify limits of overfilling (with the exception of textiles), which is usually controlled by the packer for economic, compliance, and other reasons.
(Amended 2010)

(4) Maximum Allowable Variation

The limit of the "reasonable minus variation" for an underfilled package is called a "Maximum Allowable Variation" (MAV). An MAV is a deviation from the labeled weight, measure, or count of an individual package beyond which the deficiency is considered an unreasonable minus error. Each sampling plan limits the number of negative package errors permitted to be greater than the MAV.
(Amended 2010)

(5) Deviations Caused by Moisture Loss or Gain

Deviations from the net quantity of contents caused by the loss or gain of moisture from the package are permitted when they are caused by ordinary and customary exposure to conditions that normally occur in good distribution practice and that unavoidably result in change of weight or measure. According to regulations adopted by the U.S. Environmental Protection Agency, no moisture loss is recognized on pesticides. (See Code of Federal Regulations 40 CFR Part 156.10.)

a. **Why and when do we allow for moisture loss or gain**?

Some packaged products may lose or gain moisture and, therefore, lose or gain weight or volume after packaging. The amount of moisture loss depends upon the nature of the product, the packaging material, the length of time it is in distribution, environmental conditions, and other factors. Moisture loss may occur even when manufacturers follow good distribution practices. Loss of weight "due to exposure" may include solvent evaporation, not just loss of water. For loss or gain of moisture, the moisture allowances may be applied before or after the package errors are determined.

To apply an allowance before determining package errors, adjust the Nominal Gross Weight (see Section 2.3.6. "Determine Nominal Gross Weight and Package Errors for Tare Sample"), so the package errors are increased by an amount equal to the moisture allowance. This approach is used to account for moisture loss in both the average and individual package errors.

It is also permissible to apply the moisture allowances after individual package errors and average errors are determined.

> **Example:** *A sample of a product that could be subject to moisture loss might fail because the average error is minus or the error in several of the sample packages are found to be unreasonable errors (i.e., the package error is greater than the Maximum Allowable Variation (MAV) permitted for the package's labeled quantity).*

You may apply a moisture allowance after determining the package errors by adding the allowance to the Sample Error Limit (SEL) and then, comparing the average error to the SEL to determine compliance. The moisture allowance must be added to the MAV before evaluating sample errors to identify unreasonable minus errors.
(Amended 2010)

This handbook provides "moisture allowances" for some meat and poultry products, flour, and dry pet food. (See Chapter 2, Table 2-3. "Moisture Allowances") These allowances are based on the premise that when the average net weight of a sample is found to be less than the labeled weight, but not by an amount that exceeds the allowable limit, either the lot is declared to be within the moisture allowance or more information must be collected before deciding lot compliance or noncompliance.

Test procedures for flour, some meat, and poultry are based on the concept of a "moisture allowance" also known as a "gray area" or "no decision" area (see Section 2.3.9. "Calculations"). When the average net weight of a sample is found to be less than the labeled weight, but not more than the boundary of the "gray area," the lot is said to be in the "gray" or "no decision" area. The gray area is not a tolerance. More information must be collected before lot compliance or noncompliance can be decided. Appropriate enforcement should be taken on packages found short weight and outside of the "moisture allowance" or "gray area."
(Amended 2002)

(1) Exceptions to the Average and Individual Package Requirements

There is an exemption from the average requirement for packages labeled by count with 50 items or fewer. The reason for this exemption is that the package count does not follow a "normal" distribution even if the package is designed to hold the maximum count indicated by the label declaration (e.g., egg cartons and packages of chewing gum). Another exception permits an "allowable difference" in the capacity of glass tumblers and stemware because mold capacity doesn't follow a normal distribution.

1.3. Sampling Plans

This handbook contains two sampling plans used to inspect packages: "Category A" and "Category B." Use the "Category B" Sampling Plans to test meat and poultry products at point-of-pack locations that are subject to U.S. Department of Agriculture Food Safety and Inspection Service (FSIS) requirements. When testing all other packages, use the "Category A" Sampling Plan.

a. Why is sampling used to test packages?

Inspections by weights and measures officials must provide the public with the greatest benefit at the lowest possible cost. Sampling reduces the time to inspect a lot of packages, so a greater number of items can be inspected. Net content inspection, using sampling plans for marketplace surveillance, protects consumers who cannot verify the net quantity of contents. This ensures fair trade practices and maintains a competitive marketplace. It also encourages manufacturers, distributors, and retailers to follow good manufacturing and distribution practices.

b. Why is the test acceptance criteria statistically corrected and what are the confidence levels of the sampling plans?

Testing a "sample" of packages from a lot instead of every package is efficient, but the test results have a "sampling variability" that must be corrected before determining if the lot passes or fails. The "Category A" sampling plans give acceptable lots a 97.5 % probability of passing. An "acceptable" lot is defined as one in which the "average" net quantity of contents of the packages equals or exceeds the labeled quantity. The "Category B" sampling plans give acceptable lots at least a 50 % probability of passing. The sampling plans used in this handbook are statistically valid. That means the test acceptance criteria are statistically adjusted, so they are both valid and legally defensible. This handbook does not discuss the statistical basis, risk factors, or provide the operating characteristic curves for the sampling plans. For information on these subjects, see explanations on "acceptance sampling" in statistical reference books.

c. Why use random samples?

A randomly selected sample is necessary to ensure statistical validity and reliable data. This is accomplished by using random numbers to determine which packages are chosen for inspection. Improper collection of sample packages can lead to bias and unreliable results.

d. May audit tests and other shortcuts be used to identify potentially violative lots?

Shortcuts may be used to speed the process of detecting possible net content violations. These audit procedures may include:

- using smaller sample sizes;
- using tare lists provided by manufacturers to spot check; or
- selecting samples without collecting a random sample.

These and other shortcuts allow spot checking of more products than is possible with the more structured techniques, but do not take the place of "Category A" or "Category B" testing.

e. Can audit tests and other shortcuts be used to take enforcement action?

No. Do not take enforcement action using audit test results.

If, after an audit test, there is suspicion that the package lot is not in compliance, use the appropriate "Category A" or "Category B" sampling plan to determine if the lot complies with the package requirements.

1.4. Other Regulatory Agencies Responsible for Package Regulations and Applicable Requirements

In the United States, several federal agencies issue regulations regarding package labeling and net contents. The U.S. Department of Agriculture (USDA) regulates meat and poultry. The Food and Drug Administration (FDA) regulates food, drugs, cosmetic products, and medical devices under the Food, Drug, and Cosmetic Act (FDCA) and the Fair Packaging and Labeling Act (FPLA). The Federal Trade Commission (FTC) regulates most non-food consumer packaged products as part of the agency's responsibility under the FPLA. The Environmental Protection Agency (EPA) regulates pesticides. The Bureau of Alcohol and Tobacco Tax and Trade Bureau (TTB) in the U.S. Department of the Treasury promulgates regulations for packaged tobacco and alcoholic beverages as part of its responsibility under the Federal Alcohol Administration Act.

Packaged goods produced for distribution and sale also come under the jurisdiction of state and local weights and measures agencies that adopt their own legal requirements for packaged goods. Federal statutes set requirements that pre-empt state and local regulations that are or may be less stringent or not identical to federal regulation depending on the federal law that authorizes the federal regulation. The application of Handbook 133 procedures occurs in the context of the concurrent jurisdiction among federal, state, and local authorities. Therefore, all agencies using this handbook should keep abreast of the revisions to federal agency regulations that may contain sampling or testing information not in the regulations at the time of publication of this handbook. (See Appendix A, Table 1-1. "Agencies Responsible for Package Regulations and Applicable Requirements" for information on the responsible agencies for package regulations. The requirements of this handbook must be used when testing products concurrently subject to pre-emptive federal regulations.)

1.5. Assistance in Testing Operations

If the storage, display, or location of any lot of packages requires special equipment or an abnormal amount of labor for inspection, the owner or the operator of the business must supply the equipment and/or labor as required by the weights and measures official.

1.6. Health and Safety

This handbook cannot address all of the health and safety issues associated with its use. The inspector is responsible for determining the appropriate safety and health practices and procedures before starting an inspection (e.g., contact the establishment's health and safety official). Comply with all handling, health, and safety warnings on package labels and those contained in any associated material safety data sheets

(MSDS). The inspector must also comply with federal, state, and local health and safety laws, and other appropriate requirements in effect at the time and location of the inspection. Contact your supervisor to obtain information regarding your agency's health and safety policies and to obtain appropriate safety equipment.

1.7.   Good Measurement Practices

The procedures in this handbook are designed to be technically sound and represent good measurement practices. To assist in documenting tests, we have included "model" inspection report forms designed to record the information.

(1)   Traceability Requirements for Measurement Standards and Test Equipment

Each test procedure presented in this handbook includes a list of the equipment needed to perform the inspection. The scales and other measurement standards used (e.g., balances, mass standards, volumetric, and linear measures) to conduct any test must be traceable to the National Institute of Standards and Technology (NIST). Standards must be used in the manner for which they were designed and calibrated.

(2)   Certification Requirements for Standards and Test Equipment

All measurement standards and test equipment identified in this handbook or associated with the test procedures must be calibrated or standardized before initial use. This must be done according to the calibration procedures and other instructions found on NIST's Laboratory Metrology and Calibration Procedures website at http://www.nist.gov/pml/wmd/labmetrology/calibration.cfm or using other recognized procedures (e.g., those adopted for use by a state weights and measures laboratory). After initial certification, the standards must be routinely recertified according to your agency's measurement assurance policies.

THIS PAGE INTENTIONALLY LEFT BLANK

Chapter 2. Basic Test Procedure – Gravimetric Testing

2.1. Gravimetric Test Procedure for Checking the Net Contents of Packaged Goods

The gravimetric test method uses weight measurement to determine the net quantity of contents of packaged goods. This handbook includes general test methods to determine the net quantity of contents of packages labeled in terms of weight and special test methods for packages labeled in terms of fluid measure or count. Gravimetric testing is the preferred method of testing most products because it reduces destructive testing and improves measurement accuracy.

2.2. Measurement Standards and Test Equipment

a. What type of scale is required to perform the gravimetric test method ?

Use a scale (for this handbook the term "scale" includes balances) that has at least 100 scale divisions. It must have a load-receiving element of sufficient size and capacity to hold the packages during weighing. It also requires a scale division no larger than $\frac{1}{6}$ of the Maximum Allowable Variation (MAV) for the package size being weighed. The MAV/6 requirement ensures that the scale has adequate resolution to determine the net contents of the packages. Subsequent references to product test results requiring the agreement to within one scale division based on scale divisions that are equal to or only slightly smaller than the MAV/6. (See Appendix A, Table 2-5. "Maximum Allowable Variations (MAVs) for Packages Labeled by Weight.")

> **Example:** *The MAV for packages labeled with a net weight 113 g (0.25 lb) is 7.2 g (0.016 lb). Divide ($\div$) the MAV by 6 to obtain the maximum scale division that can be used to determine the gross, tare and net weights for a package size.*
>
> $$7.2 \text{ g } (0.016) \div 6 = 1.2 \text{ g } (0.002 \text{ lb})$$
>
> *In this example, a 1 g (0.002 lb) scale division would be the maximum scale division appropriate for weighing these packages.*

(Amended 2010)

b. How often should I verify the accuracy of a scale?

Verify the accuracy of a scale before each initial daily use, each use at a new location, or when there is any indication of abnormal equipment performance (e.g., erratic indications). Recheck the scale accuracy if it is found that the lot does not pass, so there can be confidence that the test equipment is not at fault.

c. Which accuracy requirements apply?

Scales used to check packages must meet the acceptance tolerances specified for their accuracy class in the current edition of NIST Handbook 44 (HB 44) "Specifications, Tolerances, and Other Technical

Requirements for Weighing and Measuring Devices." The tolerances for Class II and Class III scales are presented in HB 44, Chapter 2.20. "Scales- T.N. Tolerances Applicable to Devices Marked I, II, III, III L, and IIII."

Note: If the package checking scale is not marked with a "class" designation, use Table 2-1. "Class of Scale" to determine the applicable tolerance.

d. What considerations affect measurement accuracy?

Always use good weighing and measuring practices. For example, be sure to use weighing and measuring equipment according to the manufacturer's instructions and make sure the environment is suitable. Place scales and other measuring equipment (e.g., flasks and volumetric measures) on a rigid support and maintain them in a level condition if being level is a requirement to ensure accuracy.

e. In testing, which tolerances apply to the scale?

Do not use a scale if it has an error that exceeds the specified tolerance in any of the performance tests described in the following section.

Steps:

1. Determine the total number of divisions (i.e., the minimum increment or graduation indicated by the scale) of the scale by dividing the scale's capacity by the minimum division.

    Example: *A scale with a capacity of 5000 g and a minimum division of 0.1 g has 50 000 divisions.*

    $$5000 \div 0.1 \, g = 50\,000 \, division$$

2. From Table 2-1. "Class of Scale", determine the class of the scale using the minimum scale division and the maximum number of scale divisions.

    Example: *On a scale with a minimum division of 0.1 g and 50 000 total scale divisions the appropriate class is "II."*

Note: If a scale is used where the number of scale divisions is between 5001 and 10 000 and the division size is 0.1 g or greater and is not marked with an accuracy Class II marking, Class III scale tolerances apply.

3. Determine the tolerance from Table 2-2. "Acceptance Tolerances for Class of Scale Based on Test Load in Divisions" in divisions appropriate for the test load and class of scale.

Steps:

Example: *Determine the number of divisions for any test load by dividing the value of the mass standard being applied by the minimum division indicated by the scale. For example, if the scale has a minimum division of 0.1 g and a 1500 g mass standard is applied, the test load is equal to 15 000 divisions (1500/0.1). On a Class II scale with a test load between 5001 and 20 000 divisions, Table 2-2. "Acceptance Tolerances for Class of Scale Based on Test Load in Divisions" indicates the tolerance is plus or minus 1 division.*

Table 2-1. Class of Scale			
Value of Scale Division[1]	**Minimum and Maximum Number of Divisions**		**Class of Scale**
	Minimum	**Maximum**	
1 mg to 0.05 g	100	100 000	II
0.1 g or more	5 000	100 000	II
0.1 g to 2 g 0.000 2 lb to 0.005 lb 0.005 oz to 0.125 oz	100	10 000	III
5 g or more 0.01 lb or more 0.25 oz or more	500	10 000	III

[1]On some scales, manufacturers designated and marked the scale with a verification division (e) for testing purposes (e = 1 g and d = 0.1 g). For scales marked Class II, the verification division is larger than the minimum displayed division. The minimum displayed division must be differentiated from the verification scale division by an auxiliary reading means such as a vernier, rider, or at least a significant digit that is differentiated by size, shape, or color. Where the verification division is less than or equal to the minimum division, use the verification division instead of the minimum division. Where scales are made for use with mass standards (e.g., an equal arm balance without graduations on the indicator), the smallest mass standard used for the measurement is the minimum division.

Table 2-2. Acceptance Tolerances for Class of Scale Based on Test Load in Divisions		
Test Load in Divisions		**Tolerance**
Class II Scale	**Class III Scale**	
0 to 5000	0 to 500	Plus or Minus 0.5 Division
5001 to 20 000	501 to 2 000	Plus or Minus 1 Division
20 001 or more	2001 to 4000	Plus or Minus 1.5 Divisions
Not Applicable	4001 or more	Plus or Minus 2.5 Divisions

f. Which performance tests should be conducted to ensure the accuracy of a scale?

Use the following procedures to verify the scale. These procedures, which are based on those required in NIST Handbook 44, have been modified to reduce the amount of time required for testing scales in field situations.

(1) Increasing-Load Test

Use certified mass standards to conduct an "increasing-load test" with all test loads centered on the load-receiving element. Start the test with the device on zero and progress with increasing test loads to a "maximum test load" of at least 10 % more than the gross weight of the packages to be tested. Use at least three different test loads of approximately equal value to test the device up to the "maximum test load." Verify the accuracy of the device at each test load. Include the package tare weight as one of the test points.

(2) Decreasing-Load Test

For all types of scales, other than one with a beam indicator or equal-arm balance, conduct a "decreasing-load test" with all test loads centered on the load-receiving element. Use the same test loads used in the "increasing-load test" of this section, and start at the "maximum test load." Remove the test loads in the reverse order of the increasing-load test until all test loads are removed. Verify the accuracy of the scale at each test load.

(3) Shift Test

When conducting a Shift Test on Bench Scales or Balances, use a test load equal to one-third of the "maximum test load" used for the "increasing-load test." For bench scales (see Figure 1-1. "Bench Scales or Balances") apply the test load as nearly as possible at the center of each quadrant of the load receiving element as shown in Figure 2-1. "Bench Scale or Balances."

For Equal Arm Balances, use a test load equal to one-half capacity centered successively at four points positioned equidistance between the center and the front, left, back, and right edges of each pan as shown (see Figure 2-2. "Equal-Arm Balance)." For example, where the load-receiving element is a rectangular or circular shape, place the test load in the center of the area represented by the shaded.

Figure 2-1. Bench Scales or Balances   **Figure 2-2. Equal-Arm Balance**

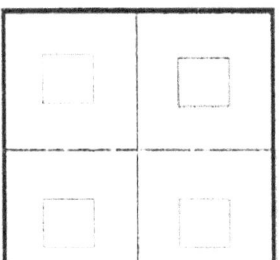   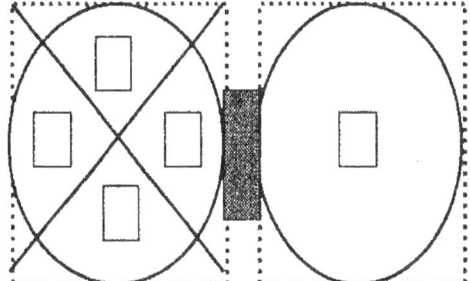

(Amended 2010)

(4) Return to Zero

Conduct the return to zero test whenever all the test weights from the scale are removed; check to ensure that it returns to a zero indication.

g. Which standards apply to other test equipment?

Specifications, tolerances, and other technical requirements for the other measurement standards and test equipment cited in this handbook are specified in the following NIST publications. These publications may be obtained from the Weights and Measures Division (http://www.nist.gov/pml/wmd/).

- Mass Standards – Use NIST Handbook 105-1, "Specifications and Tolerances for Reference Standards and Field Standard Weights and Measures – Field Standard Weights (NIST Class F)" (1990)

- Volumetric Flasks and Cylinders – Use NIST Handbook 105-2, "Specifications and Tolerances for Reference Standards and Field Standard Weights and Measures – Field Standard Measuring Flasks" (1996)

- Stopwatches – Use NIST Handbook 105-5, "Specifications and Tolerances for Reference Standards and Field Standard Weights and Measures – Field Standard Stopwatches" (1997)

- Thermometers – Use NIST Handbook 105-6, "Specifications and Tolerances for Reference Standards and Field Standard Weights and Measures – Specifications and Tolerances for Thermometers" (1997)

2.3. Basic Test Procedure

The following steps apply when gravimetrically testing any type of packaged product except Borax and glazed or frozen foods. If the tested products contain Borax, refer to Section 2.4, "Borax." If encased-in-ice or ice glazed food is tested, refer to Section 2.6, "Determining the Net Weight of Encased-in-Ice and Ice Glazed Products."

Steps:

1. Identify and define the inspection lot.

2. Select the sampling plan.

3. Select the random sample.

4. Measure the net contents of the packages in the sample.

5. Evaluate compliance with the Maximum Allowable Variation (MAV) requirement.

6. Evaluate compliance with the average requirement.

2.3.1. Define the Inspection Lot

The official defines which packages are to be tested and the size of the inspection lot. The lot may be smaller or larger than the production lot defined by the packer. Only take action on the packages contained in the lot that has been defined.

Note: Normally, there will never be access to the entire "production lot" from a manufacturer. The "inspection lot" is selected from packages that are available for inspection/test at any location in the distribution chain.

> **Example:** *An inspection lot should consist of all of the cans of a single brand of peach halves, labeled with a net quantity of 453 g (1 lb). When packages are tested in retail stores, it is not necessary to sort by lot code. If lot codes are mixed during retail testing, be sure to record the lot codes for all of the packages included in the sample so that the inspector and other interested parties can follow up on the information. For special reasons, such as a large number of packages or the prior history of problems with the product or store, the inspector may choose to define a lot as only one type of packaged product (e.g., ground beef). Another reason to narrowly define the lot is if the results of an audit test indicate the possibility of a shortage in one particular lot code within a particular product.*

a. **What is the difference between standard and random weight packages?**

Standard packages are those with identical net content declarations such as containers of soda in 2 L bottles and 2.26 kg (5 lb) packages of flour. "Random packages" are those with differing or no fixed pattern of weight, such as packages of meat, poultry, fish, or cheese.

2.3.2. Select Sampling Plans

This handbook contains two sampling plans used to inspect packages: "Category A" and "Category B." Use the "Category B" Sampling Plans to test meat and poultry products at point-of-pack locations that are subject to U.S. Department of Agriculture Food Safety and Inspection Service (FSIS) requirements. When testing all other packages, use the "Category A" Sampling Plan.

a. **Where are sampling plans located for "Category A" inspections?**

Use Appendix A, Table 2-1. "Sampling Plans for Category A," to conduct "Category A" inspections.

b. **Where are sampling plans located for "Category B" inspections?**

Use Appendix A, Table 2-2. "Sampling Plans for Category B," to conduct "Category B" inspections.

2.3.3. Basic Recordkeeping

a. **How are the specific steps of the Basic Test Procedure documented?**

Use an official inspection report to record the information. Attach additional worksheets, test notes, and other information as needed. This handbook provides random and standard packaged products model inspection report forms in Appendix C, "Model Inspection Report Forms." (Refer to Appendix C for instructions on how to complete the forms' box numbers.) Modify the model reports and the box numbers to meet your agency's needs. Other formats that contain more or less information may be acceptable.

Note: Inspection reports should be legible and complete. Good recordkeeping practices typically include record retention for a specified period of time.

Steps:

1. Record the product identity, packaging description, lot code, location of test, and other pertinent data.

2. Record the labeled net quantity of contents in Box 1. Record both metric and inch-pound declarations if they are provided on the package label.

   Example: *If the labeled weight is 453 g (1 lb), record this in Box 1.*

3. When the declaration of net quantity on the package includes both the International System of Units (SI) (metric) and inch-pound units, the larger of the two declarations must be verified. The rounding rules in NIST Handbook 130, "Uniform Packaging and Labeling Regulations" permit packers to round declarations up or down based on their knowledge of their package filling targets and the accuracy of packaging equipment.

   Determine the larger of the values by converting the SI declaration to inch-pound units, or vice versa, using conversion factors that are accurate to at least six places. Compare the values, and use the larger value in computing the nominal gross weight (see later steps). Indicate on the report which of the declarations is being verified when packages labeled with two units of measure are encountered.

   Example: *If the net weight declared on a package is 1 lb, the metric equivalent (accurate to six significant digits) is 453.592 g. Do not round*

15

Steps:

down or truncate values in the calculations until the nominal gross weight is determined and recorded. If the package is also labeled 454 g, then the metric declaration is larger than the inch-pound declaration and should be used to verify the net contents of the package. The Basic Test Procedure does not prohibit the use of units of weight instead of dimensionless units when recording package errors, nor does it prohibit the use of net content computer programs to determine product compliance.

4. Record the unit of measure in Box 2. The unit of measure is the minimum division of the unit of measurement used to conduct the test. If a scale is used that reads to thousandths of a pound, the unit of measure is 0.001 lb even if the scale division is 0.002 lb or 0.005 lb.

Examples: *If the scale has a scale division of 0.5 g, the unit of measure is 0.1 g. If a weighed package that has an error of "−0.5 g," record the error as "−5" using "dimensionless units."*

$$-0.5 \ g \div 0.1 = 5 \ dimensionless \ units$$

If the scale indicates in increments of 0.002 lb, the unit of measure is 0.001 lb. If a weighed package has an error of "+0.016," record the error as "+16" using "dimensionless units."

$$0.016 \div 0.001 = 16 \ dimensionless \ units$$

When using dimensionless units, multiply package errors by the unit of measure to obtain the package error in weight.

5. Enter the appropriate MAV value in Box 3 for the type of package (weight, volume, etc.), the labeled net contents, and the unit of measure.

b. **Where are Maximum Allowable Variations found?**

Find the MAV values for packages labeled by weight, volume, count, and measure in the tables listed below in Appendix A, Tables.

Maximum Allowable Variations

- packages labeled by weight   See Table 2-5.

- packages labeled by volume, liquid or dry   See Table 2-6.

- packages labeled by count   See Table 2-7.

- packages labeled by length, (width), or area   See Table 2-8.

- packages bearing a USDA seal of inspection – Meat and Poultry   See Table 2-9.

- textiles, polyethylene sheeting and film, mulch and soil labeled by volume,   See Table 2-10. packaged firewood, and packages labeled by count with 50 items or fewer, and specific agriculture seeds labeled by count.

(Amended 2010)

c. How is the value of an MAV found?

Refer to the appropriate table of MAVs and locate the declared quantity that is on the package label in the column marked "Labeled Quantity."   Read across the table to find the value in the column titled "Maximum Allowable Variation." Record this number in Box 3. Determine the MAV in dimensionless units and record in Box 4 on the Standard Package Report Form (a dimensionless unit is obtained by dividing the MAV recorded in Box 3 by the unit of measure recorded in Box 2). Refer to Appendix F. "Glossary," for the definition of dimensionless units.

d. How many unreasonable minus errors (UMEs) are permitted in a sample?

To find out how many minus package errors are permitted to exceed the MAV, (errors known as unreasonable minus errors or UMEs), see Column 4 in either Table 2-1. "Sampling Plans for Category A" or Table 2-2. "Sampling Plans for Category B." (refer to Appendix A) Record this number in Box 8.

(Amended 2010)

2.3.4. Random Sample Selection

a. How are sample packages selected?

Randomly select a sample from the inspection lot. Random number tables (see Appendix B. "Random Number Tables") or a calculator that is able to generate random numbers may be used to identify the sample. If the packages for the sample are not randomly selected, the test results may not be statistically valid.

Note: If the inspector and the party that is ultimately responsible for the packing and declaration of net weight for the product agree to an alternative method of sample selection, document how the sample packages were selected as part of the inspection record.

b. How is the size of the "Lot" determined?

Count the number of packages comprising the inspection lot or estimate the size to within 5 % and record the inspection lot size in Box 5.

c. How is the sample size determined?

Refer to Appendix A. Table 2-1. "Sampling Plans for Category A" or Table 2-2. "Sampling Plans for Category B" to determine the sample size. In Column 1, find the size of the inspection lot (the number recorded in Box 5 of the report form). Read across from Column 1 to find the appropriate sample size in Column 2 and record this number in Box 6 of the report form.

2.3.5. Measure Net Contents/Tare Procedures

a. What types of tare may be used to determine the net weight of package goods?

This handbook defines three types of tare for the inspection of packaged goods. The tare weight may vary considerably from package to package as compared with the variability of the package net contents, even for packages in the same production lot. Although this is not common for most packaging, the basic test procedure in this handbook considers the variation for all tare materials.

(1) Used Dry Tare

Used dry tare is used tare material that has been air dried, or dried in some manner to simulate the unused tare weight. It includes all packaging materials that can be separated from the packaged product, either readily (e.g., by shaking) or by washing, scraping, ambient air drying, or other techniques involving more than "normal" household recovery procedures, but not including laboratory procedures like oven drying. Labels, wire closures, staples, prizes, decorations, and such are considered tare. Used Dry Tare is available regardless of where the packages are tested. The net content verification procedures described in this handbook reference Used Dry Tare.

Note: When testing frozen foods with Used Dry Tare, the frost found inside frozen food packages is included as part of the net contents, except in instances in which glazed or frozen foods are tested according to Section 2.6. "Determining the Net Weight of Encased-in-Ice and Ice Glazed Products."

(2) Unused Dry Tare

Unused dry tare is all unused packaging materials (including glue, labels, ties, etc.) that contain or enclose a product. It includes prizes, gifts, coupons, or decorations that are not part of the product. If testing packages in retail store locations where they are packaged, and sold in small quantities to the ultimate consumers, the basic test procedure may be modified by using samples of the packaging material available in the store.

(3) Wet Tare

Wet Tare is used tare material where no effort is made to dry the tare material. Free-flowing liquids are considered part of the tare weight.

Wet tare procedures must not be used to verify the labeled net weight of packages of meat and poultry packed at an official United States Department of Agriculture (USDA) facility and bearing a USDA seal of inspection. The USDA Food Safety and Inspection Service (FSIS) adopted specific sections of the 2005 4th Edition of NIST HB 133 by reference in 2008 but not the "wet tare" method for determining net weight compliance. FSIS considers the free-flowing liquids in packages of meat and poultry products, including single-ingredient, raw poultry products, to be integral components of these products (see Federal Register, September 9, 2008 [Volume 73, Number 175] [Final Rule – pages 52189-52193]).

If the jurisdiction uses wet tare to determine net weight, follow the procedures described below that reference Used Dry Tare, except make no effort to dry the tare material. If Wet Tare is used to verify the

net weight of the packages, the inspector must allow for moisture loss.
(Amended 2010)

b. **How is a tare weight determined?**

Except in the instance of applying unused dry tare, select the packages for the initial tare sample from the sample packages. Mark the first two (three or five) packages in the order the random numbers were selected; these packages provide the initial tare sample. Determine the gross weight of each package and record it in Block a, "Gross Wt," under the headings "Pkg. 1," "Pkg. 2," "Pkg. 3," etc. on the report form. Except for aerosol or other pressurized packages, open the sample packages, empty, clean, and dry them as appropriate for the packaging material.

c. **How is the number of packages in the initial tare sample determined?**

To determine the initial tare sample size, see Column 5 under initial tare sample size in Appendix A, Table 2-1. "Sampling Plans for Category A" or Column 3 under initial tare sample size in Appendix A, Table 2-2. "Sampling Plans for Category B." Record the initial tare sample size in Box 7 on the report form.

Note: The initial tare sample size is considered the total tare sample size for the inspection lot when the sample size is less than 12.

d. **How is the total number of packages to be opened for tare determined and the tare weight of the packaging material determined?**

Steps:

1. Except for unused dry tare at the point-of-pack, first determine the tare weight for each package in the initial tare sample and record the value in Row b, "Tare Wt." under the appropriate package number column.

2. For sample sizes of 12 or more, subtract the individual tare weights from the respective package gross weights (Block a, minus Block b, on the report form) to obtain the net weight for each package and record each value in Block c, "Net Wt.," on the report form.

   Determine and record the "range of package errors" (called R_c) for the initial tare sample in Box 9 on the report form. (The range is the difference between the package errors.)
   (Amended 2002)

3. Determine and record the "range of tare weights" (called R_t) in Box 10.

4. Compute the ratio R_c/R_t by dividing the value in Box 9 by the value in Box 10. Record the resulting value in Box 11. (R_c and R_t must both be in the same unit of measure or both in dimensionless units.)

5. Determine and record in Box 12 the total number of packages to be opened for the tare determination from either Appendix A. Table 2-3. "Category A" or Table 2-4. "Category B."

Steps:

> ➢ In the first column (titled Ratio of R_c/R_t), locate the range in which the computed R_c/R_t falls. Then, read across to the column headed with the appropriate sample size.

> ➢ If the total number of packages to open equals the number already opened, go to step 6.

> ➢ If the total number of packages to open is greater than the number of packages already opened, compute the number of additional packages to open for the tare determination open and weigh as per step 1 and step 2 and go to step 6. Enter the total number of tare samples in Box 12.

6. Determine the average tare weight using the tare weight values for all the packages opened and record the average tare weight in Box 13.

e. **Does the inspection of aerosol containers require special procedures?**

Yes, aerosol containers are handled differently for two reasons. First, regulations in NIST HB 130 under the Uniform Packaging and Labeling Regulation (UPLR) require that packages designed "to deliver" the product under pressure, "must state the net quantity of the contents that will be expelled when the instructions for use as shown on the container are followed." This means that any product retained in aerosol containers after full dispersion is included in the tare weight. Second, aerosol containers must not be opened because they are pressurized; for safety reasons they should not be punctured or opened. When emptying aerosol containers to determine a tare weight, exhaust them in a well-ventilated area (e.g., under an exhaust hood or outdoors) at least 15 m (50 ft) from any source of open flame or spark.

To ensure that the container properly dispenses the product, read and follow any dispensing instructions on the package. If shaking during use is specified in the instructions, periodically shake (at least two or three times during expulsion of the product). If directions are not given, shake the container five times with a brisk wrist twisting motion. If the container has a ball agitator, continue the shaking procedure for one minute after the ball has shaken loose.

f. **How is the tare of vacuum-packed coffee determined?**

The gross weight of a can of vacuum-packed coffee will be more after the seal is broken and air enters the can. In the procedure to determine the tare weight of the packaging material, correct the gross weight determined for unopened cans as follows. Use the initial tare sample packages, weigh, and record the gross weight of the product-filled cans before and after breaking the vacuum seal. Compute the average gross weight difference (open weight minus sealed weight) and record this in Box 13a of the report form. The nominal gross weight equals the average tare weight minus the average difference in gross weights plus the labeled weight (Box 14): Box 13 − Box 13a + Box 1.

g. **When and where is unused dry tare used, and how is it used to determine an average tare weight?**

You may determine the average tare weight using samples of unused dry tare when testing meat, poultry, or any other products that are not subject to regulation of the Food and Drug Administration (FDA). You may utilize unused dry tare samples when conducting inspections at locations where the point-of-pack and sale are identical (e.g., store-packed products in a supermarket meat case). To determine unused dry tare at the point-of-sale, randomly select two (2) samples of unused dry tare, and weigh each separately. If there is no measurable variation in weight between the samples, proceed with the test using the weight of one of the samples. If the weight of the two (2) initial samples varies, randomly select three (3) additional tare samples and determine the average weight of all five (5) samples. Use this value as the average tare weight.

(Amended 2002)

2.3.6. Determine Nominal Gross Weight and Package Errors for Tare Sample

a. **How do I compute a nominal gross weight?**

A nominal gross weight is used to calculate package errors. To compute the nominal gross weight, add the average tare weight (recorded in Box 13) to the labeled weight (recorded in Box 1).

The nominal gross weight is represented by the formula:

Nominal gross weight = average tare + labeled weight

b. **How do I compute the package error?**

To obtain the package error, subtract the nominal gross weight from each package's gross weight. The package error is represented by the formula:

Package error = gross weight − nominal gross weight

(Added 2010)

c. **How are individual package errors determined for the tare sample packages?**

Determine the errors of the packages opened for tare by subtracting the nominal gross weight recorded in Box 14 from the individual package gross weights recorded for each package (Pkg. 1, Pkg. 2, etc.) in Block a, "Gross Wt." The nominal gross weight must be used, rather than the actual net weight, for each package to determine the package error. This ensures that the same average tare weight is used to determine the error for every package in the sample, not just the unopened packages.

- **Standard Packages.** – Record the package error in the appropriate plus or minus column on the report form for each package opened for tare.

- **Random Packages.** – Determine the package error for the tare sample using a nominal gross weight for each package so that all of the package errors are determined with the same tare weight value. Record the package error on the Random Package Report Form in the appropriate plus or minus column under Package Errors.

Note: Converting the package error to dimensionless units allows the inspector to record the package errors as whole numbers disregarding decimal points and zeroes in front and unit of measure after the number.

> **Example:** *If weighing in 0.001 lb increments, the unit of measure is 0.001 lb. If the package error for the first package opened for tare is +0.008 lb, instead of recording 0.008 lb in the plus column, record the error as "8" in the plus column. If the second package error is +0.060 lb, record the package error as "60" in the plus column, and so on. (This section does not prohibit the use of software or units of weight instead of dimensionless units.)*

d. How are individual package errors determined for the other packages in the sample?

Compare the gross weight of each of the unopened sample packages with the nominal gross weight (Box 14). Record the package errors in the "Package Errors" section of the report form using either units of weight (lb or g) or dimensionless units.

e. How is the total package error computed?

Add all the package errors for the packages in the sample. Be sure to subtract the minus package errors from the plus package errors and to record the total net error in Box 15, indicating the positive or negative value of the error.

(Amended 2010)

2.3.7. Evaluating Results

a. How is it determined if a sample passes or fails?

The following steps lead the inspector through the process to determine if a sample passes or fails. If the product is subject to moisture allowance, follow the procedures under Section 2.3.8. "Moisture Allowances" to correct the MAV.

b. How is it determined if packages exceed the Maximum Allowable Variation?

Compare each minus package error with the MAV recorded in Box 3 or Box 4 (if using dimensionless units). Circle the package errors that exceed the MAV. These are "unreasonable errors." Record the number of unreasonable minus errors found in the sample in Box 16.

c. **How is it determined if the negative package errors in the sample exceed the number of MAVs allowed for the sample?**

Compare the number in Box 16 with the number of unreasonable errors allowed (recorded in Box 8). If the number found exceeds the allowed number, the lot fails. Record in Box 17 whether the number of unreasonable errors found is less or more than allowed.

d. **How is the average error of the sample determined and does the inspected lot pass or fail the average requirement?**

Determine the average error by dividing the total error recorded in Box 15 by the sample size recorded in Box 6. Record the average error in Box 18 if using dimensionless units or in Box 19 if using units of weight. Compute the average error in terms of weight (if working in dimensionless units up to this time) by multiplying the average error in dimensionless units by the unit of measure and record the value in Box 19.

Note: If the total error recorded in Box 15 is a plus value, and Box 17 is "No," (the number of unreasonable errors is equal to or less than the number allowed, recorded in Box 8), the lot passes.

Steps:

1. If the average error is positive, the inspection lot passes the average requirement.

2. If the average error is negative, the inspection lot fails under a "Category B" test. Record in Box 20.

3. If the average error is a negative value when testing under the Sampling Plans for "Category A," compute the Sample Error Limit (SEL) as follows:

   ➤ Compute the Sample Standard Deviation and record it in Box 21.

$$s = \sqrt{\frac{1}{n-1}\sum_{i=1}^{n}(X_i - \bar{X})^2}$$

   ➤ Obtain the Sample Correction Factor from Column 3 of Appendix A. Table 2-1. "Sampling Plans for Category A" test. Record this value in Box 22.

   ➤ Compute the Sample Error Limit using the formula:

   Sample Error Limit (Box 23)
   = Sample Standard Deviation (Box 21) x Sample Correction Factor (Box 22)

4. Compliance Evaluation of the Average Error:

   ➤ If the value of the Average Error (Box 18) is smaller than the Sample Error Limit (Box 23), the inspection lot passes.

Steps:

> If the value of the Average Error (disregarding the sign) (Box 18) is larger than the Sample Error Limit (Box 23), the inspection lot fails. However, if the product is subject to moisture loss, the lot does not necessarily fail. Follow the procedures under "Moisture Allowances" in this chapter.

2.3.8. Moisture Allowances

When no predetermined allowance is found in NIST HB 133, the potential for moisture loss must be considered. Inspectors should follow their jurisdiction's guidance for making their determination on an acceptable moisture allowance.

(Added 2010)

a. **How is reasonable moisture loss allowed?**

If the product tested is subject to moisture loss, provide for the moisture allowance by following the steps listed below.

Determine the value of the moisture allowance if the product is listed below.

b. **What are the moisture allowances for flour, dry pet food, and other products?** (See Table 2-3. "Moisture Allowances")

Table 2-3. Moisture Allowances		
Verifying the labeled net weight of packages of:	**Moisture Allowance is:**	**Notes**
Flour	3 %	
Dry pet food	3 %	Dry pet food means all extruded dog and cat foods and baked treats packaged in Kraft paper bags and/or cardboard boxes with a moisture content of 13 % or less at time of pack.
Borax	See Section 2.4.	
Wet Tare Only[1]		
Fresh poultry	3 %	Fresh poultry is defined as poultry above a temperature of − 3 °C (26 °F) that yields or gives when pushed with the thumb.
Franks or hot dogs	2.5 %	

Table 2-3. Moisture Allowances		
Bacon, fresh sausage, and luncheon meats	0 %	For packages of bacon, fresh sausage, and luncheon meats, there is no moisture allowance if there is no free-flowing liquid or absorbent material in contact with the product and the package is cleaned of clinging material. Luncheon meats are any cooked sausage product, loaves, jellied products, cured products, and any sliced sandwich-style meat. This does not include whole hams, briskets, roasts, turkeys, or chickens requiring further preparation to be made into ready-to-eat sliced product. When there is no free-flowing liquid inside the package and there are no absorbent materials in contact with the product, Wet Tare and Used Dried Tare are equivalent.
[1]Wet tare procedures must not be used to verify the labeled net weight of packages of meat and poultry packed at an official United States Department of Agriculture (USDA) facility and bearing a USDA seal of inspection. The Food Safety and Inspection Service (FSIS) adopted specific sections of the 2005 4th Edition of NIST HB 133 by reference in 2008 but not the "wet tare" method for determining net weight compliance. FSIS considers the free-flowing liquids in packages of meat and poultry products, including single-ingredient, raw poultry products, to be integral components of these products (see Federal Register, September 9, 2008 [Volume 73, Number 175] [Final Rule – pages 52189-52193]).		

(Amended 2010)

c. **What moisture allowance is used with Used Dry Tare when testing packages that bear a USDA Seal of Inspection?**

There is no moisture allowance when inspecting meat and poultry from a USDA inspected plant when Used Dry Tare and "Category A" sampling plans are used.

d. **What moisture allowance is used with wet tare?**

When there is free-flowing liquid and liquid absorbed by packaging materials in contact with the product, all free-flowing liquid and the absorbed liquid is part of the wet tare.
(Added 2010)

Note: Wet tare procedures must not be used to verify the labeled net weight of packages of meat and poultry packed at an official United States Department of Agriculture (USDA) and bearing a USDA seal of inspection. The Food Safety and Inspection Service (FSIS) adopted specific sections of the 2005 4th Edition of NIST HB 133 by in 2008 reference but not the "wet tare" method for determining net weight compliance. FSIS considers the free-flowing liquid in packages of meat and poultry products, including single-ingredient, raw poultry products, to be integral components of these products (see Federal Register, September 9, 2008 [Volume 73, Number 175] [Final Rule – pages 52189-52193]).

See Table 2-3. "Moisture Allowances – Wet Tare Only."

2.3.9. Calculations

a. **How is moisture allowance computed and applied?**

To compute moisture allowance, multiply the labeled quantity by the decimal percent value of the allowance.

> **Example:**
> *Labeled net quantity of flour is 907 g (2 lb)*
> *Moisture Allowance is 3 % (0.03)*
> *Moisture Allowance = 907 g (2 lb) x 0.03 = 27 g (0.06 lb)*
> *Record this value in Box 13a.*

b. **How is a Moisture Allowance made prior to determining package errors?**

If the Moisture Allowance is known in advance (e.g., flour and dry pet food), it can be applied by adjusting the Nominal Gross Weight (NGW) used to determine the sample package errors. The Moisture Allowance (MA) in Box 13a is subtracted from the NGW to obtain an Adjusted Nominal Gross Weight (ANGW) which is entered in Box 14. The NGW is the sum of the Labeled Net Quantity of Contents (LNQC e.g., 907 g) and the Average Tare Weight (ATW) from Box 13.

> **Example**: *Use an Average Tare Weight of 14 g (0.03 lb)*
>
> *The calculation is:*
>
> *Labeled Net Quantity of Contents 907 g (2 lb) + Average Tare Weight 14 g (0.03 lb) = 921 g (2.03 lb) – Moisture Allowance 27 g (0.06 lb) = Adjusted Nominal Gross Weight of 894 g (1.97 lb)*

This result is entered in Box 14.

Package errors are determined by subtracting the Adjusted Nominal Gross Weight from the Gross Weights of the Sample Packages (GWSP).

> **Example:** *The calculation is:*
>
> *Gross Weight of the Sample Packages – Adjusted Nominal Gross Weight = Package Error*

Note: When the Nominal Gross Weight is adjusted by subtracting the Moisture Allowance value(s) the Maximum Allowable Variation(s) is not changed. This is because the errors that will be found in the sample packages have been adjusted by subtracting the Moisture Allowance (e.g., 3 %) from the Nominal Gross Weight. That increases the individual package errors by the amount of the moisture allowance (e.g., 3 %). If the value(s) of the MAV(s) were also adjusted it would result in doubling the allowance. MAV is always based on the labeled net quantity.
(Added 2010)

c. **How is a Moisture Allowance made after determining package errors**?

You can make adjustments when the value of the Moisture Allowance is determined following the test (e.g., after the sample fails or if a packer provides reasonable moisture allowance based on data obtained using a scientific method) using the following approach:

If the sample fails the Average Requirement but has no unreasonable package errors, only step 1 is used. If the sample passes the Average Requirement but fails because the sample included one or more Unreasonable Minus Errors (UMEs), only step 2 is used.

If the sample fails the Average and/or the Individual Package Requirements, both of the following steps are applied.

Steps:
1. Use the following approach to apply a Moisture Allowance to the sample after the test is completed:

   ➤ the Moisture Allowance (MA) is computed;

   Example:
   3 % x 907 g (2 lb) = 27 g (0.06 lb) and added to the Sample Error Limit (SEL)

   ➤ added to the Sample Error Limit (SEL);

   Example:
   if the SEL is 0.023 add 0.06 to obtain an Adjusted SEL of 0.083)

   ➤ the Adjusted Sample Error Limit (ASEL) is then compared to the Average Error of the Sample and:

   ➤ If the average error (disregarding sign) in Box 18 is smaller than the Adjusted Sample Error Limit, the sample passes.

   HOWEVER,

   ➤ If the average error (disregarding sign) in Box 18 is larger than the Adjusted Sample Error Limit, the sample fails.

2. To apply Moisture Allowance is to be applied to the Maximum Allowable Variation(s), the following method is recommended:

   ➤ compute Moisture Allowance (MA);

   Example: *3 % x 907 g (2 lb) = 27 g (0.06 lb)*

   ➤ add to Maximum Allowable Variation(s) (MAV) for labeled net quantity of the package to get Adjusted Maximum Allowable Variations (AMAVs);

Steps:

Example*:*

MAV for 907 g (2 lb) is 31.7 g (0.07 lb) + 27 g (0.06 lb) =
Adjusted Maximum Allowable Variation(s) (AMAV) of 58.7 g

➢ compare each minus package error to the AMAV;

➢ mark package errors that exceed the AMAV and record the number of unreasonable minus errors found in the sample; and

➢ if this number exceeds the number of unreasonable errors allowed, the sample fails.

(Added 2010)

d. **What should you do when a sample is in the moisture allowance (gray) area?**

When the average error of a lot of fresh poultry, franks or hot dogs is minus, but does not exceed the established "moisture allowance" or "gray area," contact the packer or plant management personnel to determine what information is available on the lot in question. Questions to the plant management representative may include:

• Is a quality control program in place?

• What information is available concerning the lot in question?

• If net weight checks were completed, what were the results of those checks?

• What adjustments, if any, were made to the target weight?

Note: If the plant management has data on the lot, such data may help to substantiate that the "lot" had met the net content requirements at the point of manufacture.

This handbook provides "moisture allowances" for some meat and poultry products, flour, and dry pet food. These allowances are based on the premise that when the average net weight of a sample is found to be less than the labeled weight, but not by an amount that exceeds the allowable limit, either the lot is declared to be within the moisture allowance or further investigation can be conducted.

Reasonable variations from net quantity of contents caused by the loss or gain of moisture from the package are permitted when caused by ordinary and customary exposure to conditions that occur under good distribution practices. If evidence is obtained and documented to prove that the lot was shipped from the packaging plant in a short-weight condition or was distributed under inappropriate or damaging distribution practices, appropriate enforcement action should be taken.

(Amended 2010)

2.4. Borax

a. How is it determined if the net weight labeled on packages of borax is accurate?

Use the following procedures to determine if packages of borax are labeled correctly. This procedure applies to packages of powdered or granular products consisting predominantly (more than 50 %) of borax. Such commodities are labeled by weight. Borax can lose more than 23 % of its weight due to moisture loss. However, it does not lose volume upon moisture loss, and this property makes possible a method of volume testing based on a density determination in the event that the net weight of the product does not meet the average or individual package requirements. This method may be used for audit testing to identify possible short-filling by weight at point-of-pack. Since the density of these commodities can vary at point-of-pack, further investigation is required to determine whether, such short-filling has occurred.

Test Equipment

- Metal density cup with a capacity of 550.6 mL or (1 dry pint)

- Metal density funnel with slide-gate and stand

- Scale or balance having a scale division not larger than 1 g or (0.002 lb)

- Rigid straightedge or ruler

- Pan suitable for holding overflow of density cup

Test Procedure

Steps:

1. Follow Section 2.3.1. "Define the Inspection Lot." Use a "Category A" sampling plan in the inspection; select a random sample; then use the following test procedure to determine product compliance.

2. If the lot does not comply by weight with the sampling plan requirements (either the average or individual package requirements), select the lightest package and record the net weight of this package.

3. Determine the empty weight of the density cup.

4. Place the density cup in the pan and put the funnel on top of the density cup. Close the funnel slide-gate.

5. Pour sufficient commodity into the funnel so that the density cup can be filled to overflowing.

6. Quickly remove the slide-gate from the funnel, allowing the commodity to flow into the density cup.

Steps:

7. Carefully, without agitating the density cup, remove the funnel and level off the commodity with the ruler or straightedge. Hold the ruler or straightedge at a right angle to the rim of the cup, and carefully draw it back across the top of the density cup to leave an even surface.

8. Weigh (in pounds) the filled density cup to determine gross weight. Subtract the empty density cup weight from the gross weight. This will give the net weight of the commodity.

b. **How is the volume determined?**

Steps:

1. Multiply the package net weight (in pounds) determined in step 8 above by 550.6.

2. Divide the answer just obtained (step 1) by the weight of the commodity in the density cup, determined in step 8 above. The result is the net volume of commodity in the package in milliliters.

3. Compare the net volume of the commodity in the package with the volume declared on the package. The volume declaration must not appear on the principal display panel. Instead, it will appear on the back or side of the package and may appear as:

Volume ____ mL per NIST Handbook 133

Note: 1 mL = 1 cm^3

c. **What action can be taken based on the results of the density test?**

If the net volume of commodity in the lightest package equals or exceeds the declared volume on the package, treat the lot as being in compliance based on volume and take no further action. If the net volume of borax in the lightest package is less than the declared volume on the package, further compliance testing will be necessary. Take further steps to determine if the lot was in compliance with net weight requirements at point-of-pack or was short-filled by weight. To determine this, perform a laboratory moisture loss analysis to ascertain the weight of the original borax product when it was fully hydrated; obtain additional data at the location of the packager; and/or investigate the problem with the packager of the commodity.

2.5. Determination of Drained Weight

Since the weight per unit volume of a drained product is of the same order of magnitude as that of the packaging liquid that is drained off, an "average nominal gross weight" cannot be used in checking packages of this type. The entire sample must be opened. The procedure is based upon a test method accepted by the U.S. Food and Drug Administration (FDA).

A tare sample is not needed because all the packages in the sample will be opened and measured.

The weight of the container plus drained-away liquid is determined. This weight is then subtracted from the gross weight to determine the package error.

Test Equipment

- Scales and weights recommended in Section 2.2. "Measurement Standards and Test Equipment" are suitable for the determination of drained weight.

- Sieves

  ➢ For drained weight of 1.36 kg or (3 lb) or less, one 20 cm or (8 in) No. 8 mesh U.S. Standard Series sieve, receiving pan, and cover

  HOWEVER

  ➢ For drained weight greater than 1.36 kg or (3 lb), one 30 cm or (12 in) sieve, with same specifications as above

  ➢ For canned tomatoes, a U.S. Standard test sieve with 11.2 mm ($^7/_{16}$ in) openings must be used.

- Stopwatch

(Amended 2010)

Test Procedure

Steps:

1. Follow the Section 2.3.1. "Define the Inspection Lot." Use a "Category A" or a "Category B" sampling plan in the inspection (depending on the location of test); select a random sample; then use the following test procedure to determine lot compliance.

2. Use Appendix C. "Standard Package Report." Fill out Boxes 1 through 8. Select the random sample. Determine and record on a worksheet the weight of the receiving pan.

3. Determine and record on a worksheet the gross weight of each individual package comprising the sample.

4. Pour the contents of the first package into the dry sieve with the receiving pan beneath it, incline sieve to an angle between 17° to 20° from horizontal to facilitate drainage, and allow the liquid from the product to drain into receiving pan for 2 minutes. (Do not shake or shift material on the sieve.) Remove sieve and product.

5. Weigh the receiving pan, liquid, wet container, and any other tare material. (Do not include sieve and product.) Record this weight as tare and receiving pan.

6. Subtract the weight of the receiving pan, determined in step 2, from the weight obtained in step 4 to obtain the package tare weight (which includes the weight of the liquid).

Steps:

7. Subtract the tare weight, found in step 6, from the corresponding package gross weight determined in step 3 to obtain the drained weight of that package. Determine the package error (drained weight – labeled drained weight).

8. Repeat steps 4 through 7 for the remaining packages in the sample, cleaning and drying the sieve and receiving pan between measurements of individual packages.

9. Transfer the individual package errors to the Standard Pack Report form.

10. To determine lot conformance, return to Section 2.3.7. "Evaluating Results."

2.6. Determining the Net Weight of Encased-in-Ice and Ice Glazed Products

a. **How should the net weight of frozen seafood, meat, poultry, or similar products encased-in-ice and frozen into blocks or solid masses be determined?**

Note: For determining the net weight of ice glazed seafood, meat, poultry, or similar products, follow the procedure in Section 2.6.b. "How should the net weight of ice glazed seafood, meat, poultry or similar products be determined?"

Test Equipment

*   Balance and weights (used to verify accuracy)

*   Partial immersion thermometer or equivalent with 1 °C (2 °F) graduations and a − 35 °C to + 50 °C (− 30 °F to +120 °F) accurate to ± 1 °C (± 2 °F)

*   Water source and hose with an approximate flow rate of 4 L to 15 L (1 gal to 4 gal) per minute for thawing blocks and other products

*   Sink or other receptacle [i.e., bucket with a capacity of approximately 15 L (4 gal)] for thawing blocks and other products

*   A wire mesh basket (e.g. used for testing large frozen blocks of shrimp) or a container that is large enough to hold the contents of one package (e.g., 2.27 kg or [5 lb] box of shrimp) and has openings small enough to retain all pieces of the product (e.g., an expanded metal test tube basket lined with standard 16-mesh screen)

*   Number 8 mesh, 20 cm (8 in) or 30 cm (12 in) sieve

*   Stopwatch

Test Procedure for Encased-in-Ice Product Only

Steps:

1. Follow Section 2.3.1. "Define the Inspection Lot." Use a "Category A" or a "Category B" sampling plan in the inspection (depending on the location of test); select a random sample; then use the following test procedure to determine lot compliance.

2. Place the unwrapped frozen seafood, meat, poultry, or similar products in the wire mesh basket or an open container to thaw (e.g. it is not placed in a plastic bag) and immerse in a 15 L (4 gal) or larger container of fresh water at a temperature between 23 °C to 29 °C (75 °F to 85 °F). Submerge the basket so that the top of the basket extends above the water level.

3. Maintain a continuous flow of water into the bottom of the container to keep the temperature within the specified range. This is accomplished by maintaining a constant flow of warm water into the container holding the product (e.g., place a bucket in a sink to catch the overflow, and feed warm water into the bottom of the bucket through a hose).

Note: Direct immersion does not result in the product absorbing moisture because the freezing process causes the tissue to lose its ability to hold water.

4. As soon as the product thaws, determined by loss of rigidity, transfer all material to a sieve (20 cm [8 in] for packages less than 453 g [1 lb] or 30 cm [12 in] for packages weighing more than 453 g [1 lb]) and distribute it evenly over the sieve.

5. Without shifting the product, incline the sieve 30° from the horizontal position to facilitate drainage, and drain for 2 minutes.

6. At the end of the drain time, immediately transfer the product to a tared pan for weighing to determine the net weight.

(Amended 2010)

b. **How should the net weight of ice glazed seafood, meat, poultry, or similar products be determined?**

For ice glazed seafood, meat, poultry or similar products, determine the net weight after removing the glaze using the following procedure.

Test Equipment

- Balance and weights (used to verify accuracy)

- Continuous cold water source

- Number 8 sieve and receiving pan, 20 cm (8 in) for packages 453 g (1 lb) or less. A 30 cm (12 in) for packages more than 453 g (1 lb).

- Means to determine a 17° to 20° angle

- Stopwatch

Test Procedures for Ice-Glazed Product Only

Steps:

1. Follow Section 2.3.1. "Define the Inspection Lot." Use a "Category A" sampling plan in the inspection; select a random sample; and use the following test procedure to determine lot compliance.

2. Fill out the header information on boxes 1 through 8 on the Ice Glazed Package Report form (See Appendix C). A tare sample is not needed. Record package price, price per pound, lot size, sample size, and unit of measure in step 1 of the Ice Glazed Package Worksheet (See Appendix C).

Note: Use an official inspection report to record the inspection information. Attach additional worksheets, test notes, and other information as needed. This handbook provides an ice glazed worksheet and package report form in Appendix C. Modify the worksheet, package report and the box numbers to meet your agency's needs. Other formats that contain more or less information may be acceptable.

3. Number each package. Weigh each package for gross package weight and enter in row 1 "Gross Package Weight" on worksheet.

4. Enter the labeled net weight in Row 2 "Labeled Net Weight" for each package on the worksheet. If dual units, determine and enter the larger of the two units.

5. Record the maximum allowable variation on row 3 "MAV" on the worksheet.

6. Weigh receiving pan and record the weight in row 4, "Receiving Pan Weight" on the worksheet.

7. Deglaze the product. Remove a package from low temperature storage; open it immediately and place the contents in the sieve or other draining device (e.g., colander) under a gentle spray of cold water. Carefully agitate the product. Handle with care to avoid breaking the product. Continue the spraying process until all ice glaze, that is seen or felt is removed. In general, the product should remain rigid; however, the ice glaze on certain products, usually smaller sized commodities, sometimes cannot be removed without partial thawing of the product. Nonetheless, remove all ice glaze, because it may be a substantial part of the package weight.

8. Transfer the product to the sieve (if the product is not already in the sieve) Without shifting the product, incline the sieve to an angle of 17 degrees to 20 degrees to facilitate drainage and drain (into waste receptacle or sink) for 2 minutes using a stopwatch.

9. At the end of the drain time immediately transfer the entire product to the receiving pan for weighing to determine the net weight.

Steps:

10. Place the product and receiving pan on the scale and weigh. Record the net weight in row 5 on the ice glazed package worksheet. The net weight of product is equal to the weight of the receiving pan and the product minus the receiving pan weight.

11. The package error is equal to the net weight of the product minus the labeled weight. Record the package error**Error! Bookmark not defined.** in row 6 on the ice glazed package worksheet.

12. Repeat steps 2 through 10 for each package in the sample, cleaning the sieve and cleaning and drying the receiving pan between package measurements.

13. Transfer data from the ice glazed package worksheet to the ice glazed package report.

Evaluation of Results

Follow the procedures in Section 2.3.7. "Evaluating Results."

(Amended 2010)

THIS PAGE INTENTIONALLY LEFT BLANK

Chapter 3. Test Procedures – For Packages Labeled by Volume

3.1. Scope

a. What types of packaged goods can be tested using these procedures?

Use this procedure to determine the net contents of packaged goods labeled in fluid volume such as milk, water, beer, oil, paint, distilled spirits, soft drinks, juices, liquid cleaning supplies or chemicals. This chapter also includes procedures for testing the capacities of containers such as paper cups, bowls, glass tumblers, and stemware.

b. What types of packages are not covered by these procedures?

These procedures do not cover berry baskets and rigid-dry measures that are covered by specific code requirements in NIST Handbook 44. "Specifications, Tolerances, and Other Technical Requirements for Weighing and Measuring Devices."

c. When can the gravimetric test procedure be used to verify the net quantity of contents of packages labeled by volume?

The gravimetric procedure may be used to verify the net quantity of contents of packages labeled in volume when the density (density means the weight of a specific volume of liquid determined at a reference temperature) of the product being tested does not vary excessively from one package to another.

d. What procedure is followed if the gravimetric test procedure cannot be used?

Test each package as described in Section 3.3. "Volumetric Test Procedure for Liquids."

e. What considerations besides density affect measurement accuracy?

In addition to possible package-to-package variations in product density, the temperature of the liquid will affect the volume of product. The product will expand or contract based on a rise or fall in product temperature.

> **Example:** *The volume of a liquid cleaning product might be 5 L (1.32 gal) at 20 °C (68 °F) and 5.12 L (1.35 gal) at 25 °C (77 °F), which represents a 2.2 % change in volume.*

Note: This extreme example is for illustrative purposes, a 2.2 % volume change will not occur in normal testing.

f. **What reference temperature should be used to determine the volume of a liquid?**

Use the reference temperature specified in Table 3-1. "Reference Temperatures for Liquids" to determine volume. When checking liquid products labeled by volume using the gravimetric procedure, maintain the packages used to determine product densities at reference temperatures. If testing the packages in a sample volumetrically, each package in the sample must be maintained at or corrected to the reference temperature when its volume is determined.

Note: When checking liquid products using a volumetric or gravimetric procedure, the temperature of the samples must be maintained at the reference temperature ± 2 °C (± 5 °F).

Table 3-1. Reference Temperatures for Liquids		
If the liquid commodity is:	**Volume is determined at the reference temperature of:**	**Code of Federal Regulation Reference***
Beer	4 °C (39.1 °F)	27 CFR, Part 7.10
Distilled Spirits	15.56 °C (60 °F)	27 CFR, Part 5.11
Frozen food - sold and consumed in the frozen state	At the frozen temperature	21 CFR §101.105(b)(2)(i)
Petroleum	15.6 °C (60 °F)	16 CFR §500.8(b)
Refrigerated food (e.g., milk and other dairy products labeled "KEEP REFRIGERATED")	4 °C (40 °F)	21 CFR §101.105(b)(2)(ii)
Other liquids and wine (e.g., includes liquids sold in a refrigerated state for immediate customer consumption such as soft-drinks, bottled water and others that do not require refrigeration)	20 °C (68 °F)	Food: 21 CFR 101.105(b)(2)(iii) Non-Food: 16 CFR §500.8(b) Wine: 27 CFR, Part 4.10 (b)
*The Code of Federal Regulations can be accessed online at: *http://www.gpoaccess.gov/*		

(Amended 2010)

3.2. Gravimetric Test Procedure for Liquids

Test Equipment

- A scale that meets the requirements in Chapter 2, Section 2.2. "Measurement Standards and Test Equipment."

Note: To verify that the scale has adequate resolution for use, it is first necessary to determine the density of the liquid; next verify that the scale division is no larger than MAV/6 for the package size under test. The smallest graduation on the scale must not exceed the weight value for MAV/6.

Example: *Assume the inspector is using a scale with 1 g (0.002 lb) increments to test packages labeled 1 L (33.8 fl oz) that have an MAV of 29 mL (1 fl oz). Also, assume the inspector finds that the weight of 1 L of the liquid is 943 g (2.078 lb). This will result in an MAV/6 value in weight of 4.715 g (0.010 lb):*

29 mL/6 = 4.8 mL	*(1 fl oz/6 = 0.166 6 fl oz)*
943 g/1000 mL = 0.943 g/mL	*(2.07 8 lb/33.6 fl oz = 0.061 8 lb/fl oz)*
4.8 mL x 0.943 g/mL = 4.5264 g	*(0.166 6 fl oz x 0.061 8 lb/fl oz = 0.010 lb)*

In this example, the 1 g (0.002 lb) scale division is smaller than the MAV/6 value of 4.5264 g (0.010 lb) so the scale is suitable for making a density determination.

- A partial immersion thermometer (or equivalent) with a range of − 35 °C to + 50 °C (30 °F to 120 °F), at least 1 °C (1 °F) graduations, and with a tolerance of ± 1 °C (± 2 °F).

- Volumetric measures

    Example: *When checking packages labeled in SI units, flask sizes of 100 mL, 200 mL, 500 mL, 1 L, 2 L, 4 L, and 5 L and a 50 mL cylindrical graduate with 1 mL divisions may be used. When checking packages labeled in inch-pound units the use of measuring flasks and graduates with capacities of gill, half-pint, pint, quart, half-gallon, gallon, and a 2 fl oz cylindrical graduate, graduated to ½ fl dr is recommended.*

- Defoaming agents may be necessary for testing liquids such as beer and soft drinks that effervesce or are carbonated. Two such products are Hexanol or Octanol (Capryl Alcohol*).

*Note: The mention of trade or brand names does not imply that these products are endorsed or recommended by the U.S. Department of Commerce over similar products commercially available from other manufacturers.

- Bubble level at least 15.24 cm (6 in) in length

- Stopwatch

Test Procedure

Steps:

1. Follow Section 2.3.1. "Define the Inspection Lot." Use a "Category A" sampling plan in the inspection. Select a random sample; then use the following procedure to determine lot compliance.

Steps:

2. Bring the sample packages and their contents to the reference temperature as specified in Table 3-1. "Reference Temperatures for Liquids." To determine if the liquid is at its reference temperature, immerse the thermometer in the liquid before starting the test. Verify the temperature again immediately after the flask and liquid is weighed. If the product requires mixing for uniformity, mix it before opening in accordance with any instructions specified on the package label. Shaking liquids, such as flavored milk, often entraps air that will affect volume measurements, so use caution when testing these products. Often, less air is entrapped if the package is gently rolled to mix the contents.

3. For milk, select a volumetric measure equal to or one size smaller than the label declaration. For all other products, select a volumetric measure that is one size smaller than the label declaration.
(Amended 2004)

> **Example:** *If testing a 1 L bottle of juice or a soft drink, select a 500 mL volumetric measure.*

Note: When determining the density of milk, if the product from the first container does not fill the volumetric measure to the nominal capacity graduation, product may be added from another container as long as product integrity is maintained (i.e., brand, identity, lot code, and temperature).

4. Prepare a clean volumetric measure to use according to the following procedures:

> ➤ Because flasks are ordinarily calibrated on a "to deliver" basis, they must be "wet down" before using. Immediately before use, fill the volumetric flask(s) or graduate with water. The water should be at the reference temperature of the product being tested. Fill the flask(s) with water to a point slightly below the top graduation on the neck. The flask should be emptied in 30 seconds (± 5 seconds). Tilt the flask gradually so the flask walls are splashed as little as possible as the flask is emptied. When the main flow stops, the flask should be nearly inverted. Hold the flask in this position for 10 seconds more and touch off the drop of water that adheres to the tip. If necessary, dry the outside of the flask. The flask or graduate is then ready to fill with liquid from a package. This is called the "wet down" condition.

Note: When using a volumetric measure that is calibrated "to contain," the measure must be dry before each measurement.

> ➤ If the liquid effervesces or foams when opened or poured (such as carbonated beverages), add two drops of a defoaming agent to the bottom of the flask before filling with the liquid. If working with a carbonated beverage, make all density determinations immediately upon placing the product into the standard. This reduces the chance of volume changes occurring from the loss of carbonization.

Steps:

5. If the flask capacity is equal to the labeled volume, pour the liquid into the volumetric measure tilting the package to a nearly vertical position. If the flask capacity is smaller than the package's labeled volume, fill the flask to its nominal capacity graduation. If conducting a volumetric test, drain the container into the flask for 1 minute after the stream of liquid breaks into drops.

6. Position the flask on a level surface at eye level. For clear liquids, place a material of some dark color outside the flask immediately below the level of the meniscus. Read the volume from the lowest point of the meniscus. For opaque liquids, read volume from the center top rim of the liquid surface.

7. Use the gravimetric procedure to determine the volume if the limit specified for the difference in density is not exceeded.

   ➢ Select a volumetric measure equal to or one size smaller than the labeled volume (depending on the product) and prepare it as described in step 4 of this section. Then determine and record its empty weight.

   ➢ Determine acceptability of the liquid density variation, using two packages selected for tare according to Section 2.3.5. "Tare Procedures" as follows:

      • Determine the gross weight of the first package.

      • Pour the liquid from the first package into a flask. Measure exactly to the nominal capacity marked on the neck of the measure.

      • Weigh the filled flask and subtract its empty weight to obtain the weight of the liquid. Determine density by dividing the weight of the liquid by the capacity of the flask.

      • Determine the weight of the liquid from a second package using the same procedure.

      • If the difference between the densities of the two packages exceeds one division, use the volumetric procedure in Section 3.3. "Volumetric Test Procedure for Liquids."

8. Determine the Average Used Dry Tare Weight of the sample according to provisions of Section 2.3.5. "Tare Procedures."

9. Calculate the Average Product Density by adding the densities of the liquid from the two packages and dividing the sum by two.

10. Calculate the "nominal gross weight" using the following formula if the flask capacity is equal to the labeled volume:

Nominal Gross Weight = (Average Product Density [in weight units]) + (Average Used Dry Tare Weight)

Steps:

Note: If the flask size is smaller than the labeled volume, the following formula is used:

Nominal Gross Weight = (Average Product Density x
[Labeled Volume/Flask Capacity]) + (Average Used Dry Tare Weight)

a. **How is "nominal gross weight" determined?**

Determine the "nominal gross weight" as follows:

Steps:

1. Determine the Average Used Dry Tare Weight of the sample according to provisions of Section 2.3.5. "Tare Procedures."

2. Calculate the Average Product Density by adding the densities of the liquid from the two packages and dividing the sum by two.

3. Calculate the "nominal gross weight" using the following formula if the flask capacity is equal to the labeled volume.

Nominal Gross Weight = (Average Product Density [in weight units]) +
(Average Used Dry Tare Weight)

Note: If the flask size is smaller than the labeled volume, the following formula is used:

Nominal Gross Weight = (Average Product Density x
[Labeled Volume/Flask Capacity]) + (Average Used Dry Tare Weight).

b. **How are the errors in the sample determined?**

Steps:
1. Weigh the remaining packages in the sample.

2. Subtract the nominal gross weight from the gross weight of each package to obtain package errors in terms of weight. All sample packages are compared to the nominal gross weight.

3. To convert the average error or package error from weight to volume, use the following formula:

Package Error in Volume = Package Error in Weight/Average Product Density Per
Volume Unit of Measure.

Evaluation of Results

Follow the procedures in Chapter 2, Section 2.3.7. "Evaluating Results" to determine lot conformance.

3.3.   Volumetric Test Procedure for Liquids

a. How is the volume of liquid contained in a package determined volumetrically?

Follow steps 1 through 6 in Section 3.2. "Gravimetric Test Procedure for Liquids" for each package in the sample.

b. How are the errors in the sample determined?

Read the package errors directly from the graduations on the measure. The reference temperature must be maintained within ± 2 °C (± 5 °F) for the entire sample.

Evaluation of Results

Follow the procedures in Chapter 2, Section 2.3.7. "Evaluating Results" to determine lot conformance.

3.4.   Other Volumetric Test Procedures

a. What other methods can be used to determine the net contents of packages labeled by volume?

Depending on how level the surface of the commodity is, use one of two headspace test procedures. Use the headspace test procedure in Section 3.4.b. to determine volume where the liquid has a level surface (e.g., oils, syrups, and other viscous liquids). Use the procedure in Section 3.4.c. to determine volume where the commodity does not have a level surface (e.g., mayonnaise and salad dressing).

Test Procedure

Before conducting any of the following volumetric test procedures follow Section 2.3.1. "Define the Inspection Lot." Use a "Category A" sampling plan in the inspection; select a random sample; then use the following procedure to determine lot compliance.

Test Equipment

- Micrometer depth gage (ends of rods fully rounded) 0 mm to 225 mm (0 in to 9 in) or longer

- Level (at least 15 cm (6 in) in length)

- Laboratory pipets and/or buret

    ➤ Class A 500 mL buret that conforms to ASTM E287-2(2007), "Standard Specification for Laboratory Glass Graduated Burets."

    ➤ Class A pipets, calibrated "to deliver" that conform to ASTM E969S-S02(2007), "Standard Specification for Glass Volumetric (Transfer) Pipets."

- Volumetric measures

- Water

- Rubber bulb syringe

- Plastic disks that are 3 mm ($^1/_8$ in) thick with diameters equal to the seat diameter or larger than the brim diameter of each container to be tested. The diameter tolerance for the disks is 50 µm (± 0.05 mm [± 0.002 in]). The outer edge should be smooth and beveled at a 30° angle with the horizontal to 800 µm (0.8 mm [$^1/_{32}$ in]) thick at the edge. Each disk must have a 20 mm (¾ in) diameter hole through its center and a series of 1.5 mm ($^1/_{16}$ in) diameter holes 25 mm (1 in) apart around the periphery of the disk and 3 mm ($^1/_8$ in) from the outer edge. All edges must be smooth.

- Stopwatch

- Partial immersion thermometer (or equivalent) with 1°C (2 °F) graduations and a range of − 35 °C to + 50 °C (− 30 °F to + 120 °F) accurate to ± 1°C (± 2 °F)

b. **How is the volume of oils, syrups, and other viscous liquids that have smooth and level surfaces determined?**

Note: Make all measurements on a level surface.

Steps:

1. Bring the temperature of both the liquid and the water to be used to measure the volume of the liquid to the reference temperature specified in Table 3-1. "Reference Temperatures for Liquids." Verify with a thermometer that product has maintained the reference temperature.

2. Measure the headspace of the package at the point of contact with the liquid using a depth gauge with a fully rounded, rather than a pointed, rod end. If necessary, support the package to prevent the bottom of the container from distorting.

3. Empty, clean, and dry the package.

4. Refill the container with water measured from a volumetric standard to the original liquid headspace level measured in step 2 of this section until the water touches the depth gauge.

5. Determine the amount of water used in step 4 of this section to obtain the volume of the liquid and calculate the "package error" based on that volume.

Evaluation of Results

Follow the procedures in Section 2.3.7. "Evaluating Results" to determine lot conformance.

c. **How is the volume of mayonnaise and salad dressing, and water immiscible products that do not have smooth and level surfaces determined?**

Use the volumetric headspace procedure described in this section to determine volume when the commodity does not have a level surface (e.g., mayonnaise, salad dressing, and other water immiscible products without a level liquid surface). The procedure guides the inspector to determine the amount of headspace above the product in the package and the volume of the container. Determine the product volume by subtracting the headspace volume from the container volume. Open every package in the sample.
(Amended 2010)

Note: Make all measurements on a level surface.

Steps:

1. Bring the temperature of both the commodity and the water used to measure the volume to the appropriate temperature designated in Table 3-1. "Reference Temperatures for Liquids."

2. Open the first package and place a disk larger than the package container opening over the opening.

3. Measurement Procedure

   ➢ Deliver water from a flask (or flasks), graduate, or buret, through the central hole in the disk onto the top of the product until the container is filled. If it appears that the contents of the flask may overfill the container, do not empty the flask. Add water until all of the air in the container has been displaced and the water begins to rise in the center hole of the disk. Stop the filling procedure when the water fills the center disk hole and domes up slightly due to the surface tension. Do not add additional water after the level of the water dome has dropped.

   ➢ If the water dome breaks on the surface of the disk, the container has been overfilled and the test is void; dry the container and start over.

4. To obtain the headspace capacity, record the volume of water used to fill the container and subtract 1mL (0.03 fl oz), which is the amount of water held in the hole in the disk specified.

5. Empty, clean, and dry the package container.

6. Using steps 3 and 4 of this section. Refill the package container with water measured from a volumetric measure to the maximum capacity of the package, subtract 1 mL (0.03 fl oz), and record the amount of water used as the container volume; and

7. From the container volume determined in step 6 of this section, subtract the headspace capacity in step 4 of this section to obtain the measured volume of the product and calculate the "package error" for that volume where "package error" equals labeled volume minus the measured volume of the product.

Evaluation of Results

Follow the procedures in Section 2.3.7 "Evaluating Results" to determine lot conformance."

3.5.   Goods Labeled by Capacity – Volumetric Test Procedure

a.   **What type of measurement equipment is needed to perform the headspace test procedures?**

Use the test equipment in Section 3.4. "Other Volumetric Test Procedures" (except for the micrometer depth gage) to perform these test procedures.

b.   **How do you determine if goods labeled by capacity meet the average and individual requirements?**

Note: Make all measurements on a level surface.

Steps:
1.   Before conducting any of the following volumetric test procedures, refer to Section 2.3.1. "Define the Inspection Lot." Use a "Category A" sampling plan in the inspection; select a random sample; then use the following test procedure to determine lot compliance.

2.   When testing goods labeled by capacity, use water at a reference temperature of 20 °C ± 2 °C (68 °F ± 5 °F).

3.   Select a sample container and place a disk larger than the container opening over the opening.

4.   Measurement Procedure

    ➤ Add water to the container using flask (or flasks), graduate, or buret corresponding to labeled capacity of the container. If it appears that the contents of the flask may overfill the container, do not empty the flask. Add water until all of the air in the container has been displaced and the water begins to rise in the center hole of the disk. Stop filling the container when the water fills the center disk hole and domes up slightly due to the surface tension.

    ➤ If the water dome breaks on the surface of the disk, the container has been overfilled and the test is void; dry the container and start over.

    ➤ Record the amount of water used to fill the container and subtract 1 mL (0.03 fl oz) (this is the amount of water held in the hole in the disk specified) to obtain the total container volume.

5.   Test the other containers in the sample according to in step 4 of this procedure.

Steps:

6. To determine package errors, subtract the total container volume obtained in steps 4 and 5 of this section from the labeled capacity of the container.

Evaluation of Results

Follow the procedures in Section 2.3.7. "Evaluating Results" to determine lot compliance.

3.6.   Pressed and Blown Glass Tumblers and Stemware

a. **What requirements apply to pressed and blown glass tumblers and stemware?**

This handbook provides a tolerance to the labeled capacity of glass tumblers and stemware. The average requirement does not apply to the capacity of these products. See Table 3-2. "Allowable Differences for Pressed and Blown Glass Tumblers and Stemware."

b. **How is it determined if tumblers and stemware meet the individual package requirement?**

Follow Section 2.3.1. "Define the Inspection Lot" and determine which sampling plan to use in the inspection, select a random sample, and then use the volumetric test procedure to determine container capacity and volume errors.

c. **What type of measuring equipment is needed to perform the test procedures?**

Use the equipment specified in Section 3.4. "Other Volumetric Test Procedures." (except for the micrometer depth gage) to perform these test procedures.

d. **What are the steps of the test procedure?**

Follow steps 1 through 7 in Section 3.5.b. "Goods Labeled by Capacity – Volumetric Test Procedure."

e. **How is it determined if the samples conform to the allowable difference?**

Compare the individual container error with the allowable difference that applies in Table 3-2. "Allowable Differences for Pressed and Blown Glass Tumblers and Stemware." If a package contains more than one container, all of the containers in the package must meet the allowable difference requirements in order for the package to pass.

Table 3-2. Allowable Differences for Pressed and Blown Glass Tumblers and Stemware	
Unit of Measure	
If the capacity in metric units is:	The allowable difference is:
200 mL or less	± 10 mL
More than 200 mL	± 5 % of the labeled capacity
If the capacity in inch-pound units is:	Then the allowable difference is:
5 fl oz or less	± ¼ fl oz
More than 5 fl oz	± 5 % of the labeled capacity

Evaluation of Results

Count the packages in the sample with volume errors greater than the allowable difference and compare the resulting number with the number given in Column 3. (Table 2-11, Appendix A)

- If the number of containers in the sample with errors exceeding the allowable difference exceeds the number allowed in Column 3, the lot fails.

- If the number of packages with errors exceeding the allowable difference is less than or equal to the number in Column 3, the lot passes.

Note: The average capacity error is not calculated because the lot passes or fails based on the individual volume errors. Act on the individual units containing errors exceeding the allowable difference individually even though the lot passes the requirement.

3.7 Volumetric Test Procedure for Paint, Varnish, and Lacquers – Non-aerosol

a. **How is the volume of paint, varnish, and lacquers contained in a package determined?**

Use one of three different test methods depending upon the required degree of accuracy and the location of the inspection. The procedures include both retail and in-plant audits, and a "possible violation" method that is designed for laboratory or in plant use because of cleanup and product collection requirements. The procedures are suitable to use with products labeled by volume and packaged in cylindrical containers with separate lids that can be resealed.

Test Equipment

- A scale that meets the requirements in Section 2.2. "Measurement Standards and Test Equipment"

- Volumetric measures

- Micrometer depth gage (ends of rods fully rounded), 0 mm to 225 mm (0 in to 9 in)

- Diameter (Pi) tape measure, 5 cm to 30 cm (2 in to 12 in)

- Spanning bar, 2.5 cm by 2.5 cm by 30 cm or (1 in by 1 in by 12 in)

- Rule, 30 cm (12 in)

- Paint solvent or other solvent suitable for the product being tested

- Cloth, 30 cm (12 in) square

- Wood, 5 cm (2 in) thick, by 15 cm (6 in) wide, by 30 cm (12 in) long

- Rubber mallet

- Metal disk, 6.4 mm (¼ in) thick and slightly smaller than the diameter of package container bottom

- Rubber spatula

- Level at least 15 cm (6 in) in length

- Micrometer (optional)

- Stopwatch

Test Procedures

Note: When instructed to record a measurement in a column, refer to the numbered columns in the "Audit Worksheet for Checking Paint" in this section.

Steps:

1. Select a random sample. A tare sample is not needed.

2. For containers less than 4 L or (1 gal):

   ➢ Measure the outside diameter of each container near its middle to the closest 0.02 mm (0.001 in).

   ➢ Use a diameter tape measure to record the measurements in Column 3.

   ➢ Place the containers on a level surface and using the micrometer depth gage, record their heights in Column 1 on the worksheet.

   ➢ If the range of outside diameters exceeds 0.125 mm (0.005 in) or the range in heights exceeds 1.58 mm (0.062 5 in), do not use this procedure. If the ranges are within the specified limits, weigh all cans in the sample, select the container with the lightest gross weight, and remove its lid. Continue with step 4 below.

Steps:

3. For 4 L (1 gal) containers:

    ➢ Gross weigh each package in the sample.

    ➢ Select the package with the lightest gross weight and remove its lid.

4. Use a direct reading diameter tape measure to measure the outside diameter of the selected container near its top, middle (already measured if step 2 was followed), and bottom to the closest 0.02 mm (0.001 in). Record these measurements in Columns 2, 3, and 4. Add the three diameter values and divide by three to obtain the average diameter and record this value in Column 5.

5. If a micrometer is available, measure the wall and the paper label thickness of the container; otherwise, assume the wall and label thicknesses given in Table 3-3. "Thickness of Paint Can Walls and Labels" below:

    Subtract twice the thickness of the wall of the can and paper label from the average can diameter (step 4) to obtain the average liquid diameter. Record the liquid diameter in Column 6.

6. On a level surface, place the container on the circular metal disk that is slightly smaller in diameter than the lower rim of the can so the bottom of the container nests on the disk to eliminate any "sag" in the bottom of the container.

7. Place the spanning bar and depth gage across the top of the paint can and mark the location of the spanning bar on the rim of the paint container. Measure the distance to the liquid level, to the nearest 20 μm (0.02 mm) (0.001 in), at three points in a straight line. Take measurements at points approximately 1 cm ($^3/_8$ in) from the inner rim for cans 12.5 cm (5 in) in diameter or less (and at 1.5 cm [$^1/_2$ in] from the rim for cans exceeding 12.5 cm [5 in]) in diameter and at the center of the can. Add the three readings and divide by three to obtain the average distance to the liquid level in the container. Record the average distance to the liquid level in Column 7.

8. Measure the distance to the bottom of the container at three points in a straight line in the same manner as outlined in step 7. Add the three readings and divide by three to obtain the average height of the container and record it in Column 8.

9. Subtract the average distance to the liquid level (Column 7) from the average height of the container (Column 8) to obtain the average height of the liquid column and record it in Column 9.

10. Determine the volume of paint in the container by using the following formula:

$$Volume = 0.7854 \, D^2 H$$

Where D = average liquid diameter (Column 6) and
H = average liquid height (Column 9)

Steps:

11. Record this value in Column 10. If the calculated volume is less than labeled volume, go to the Section 3.8.1. "Violation Procedure".

Table 3-3. Thickness of Paint Can Walls and Labels	
Can Size	**Wall Thickness**
4 L (1 gal)	250 µm (0.25 mm) [0.010 in]
2 L (½ gal)	
1 L (1 qt)	230 µm (0.23 mm) [0.009 in]
500 mL (1 pt)	
250 mL	200 µm (0.20 mm) [0.008 in]
Label Thickness* for all can sizes: 100 µm (0.10 mm) [0.004 in] (*Paper only – ignore labels lithographed directly onto the container)	

Note: Use the following format to develop worksheets to perform audits and determine the volume when checking paint. Follow the procedure and it will indicate the column in which the various measurements made can be recorded.

Example: Audit Worksheet for Checking Paint (add additional rows as needed)									
1. Can Height	Can Diameter				6. Avg. Liquid Diameter	7. Avg. Liquid Level	8. Avg. Container Depth	9. Avg. Liquid Depth	10. Volume[1]
	2. Top	3. Middle	4. Bottom	5. Average					
[1]10. Volume = 0.7854 x 6 x 6 x 9									

b. **What test procedure is used to conduct a retail audit test?**

Conduct a retail audit using the following test procedure that is suitable for checking cylindrical containers up to 4 L (1 gal) in capacity. Use step 2 in the retail audit test procedure with any size container except 4 L (1 gal), but step 3 must be used for containers with capacities of 4 L (1 gal). The method determines the volume of a single can in the sample selected as most likely to contain the smallest volume of product. Do not empty any containers because only their critical dimensions are being measured.

c. **How accurate is the dimensional test procedure?**

The configuration of the bottom of the can, paint clinging to the lid, and slight variations in the wall and label thicknesses of the paint container may produce an uncertainty estimated to be at least 0.6 % in this auditing procedure. Therefore, this method is recommended solely to eliminate from more rigorous testing those packages that appear to be full measure. Use the violation procedures when the volume determined in step 10 is less than the labeled volume or in any case where short measure is suspected.

d. **How is an in-plant audit conducted?**

Use the following procedures to conduct an in-plant audit inspection. This method applies to a container that probably contains the smallest volume of product. Duplicate the level of fill with water in a can of the same dimensions as the one under test. Use this method to check any size of package if the liquid level is within the measuring range of the depth gage. If any paint is clinging to the sidewall or lid, carefully scrape the paint into the container using a rubber spatula.

Note: When instructed to record a measurement in a column, refer to the numbered columns in the "Audit Worksheet for Checking Paint" in the Test Procedure of this section (3.7.).

Steps:
1. Follow steps 1 through 6 of the retail audit test in Section 3.7.b.

2. Place the spanning bar and depth gage across the top of the paint can. Measure the liquid level at the center of the surface and record the level in Column 7.

3. Select an empty can with the same bottom configuration as the container under test and with a diameter and height equal to that of the container under test within plus or minus the following tolerances:

    a. For 500 mL or (1 pt) cans – within 25 μm (0.025 mm) (0.001 in)
    b. For 1 L or (1 qt) cans – within 50 μm (0.05 mm) (0.002 in)
    c. For 2 L or (½ gal) cans – within 75 μm (0.075 mm) (0.003 in)
    d. For 4 L or (1 gal) cans – within 100 μm (0.1 mm) (0.004 in)

4. Set the empty can on a level work surface with a circular metal disk that is slightly smaller in diameter than the bottom can rim underneath the can to eliminate sag. Set up the spanning bar and depth gage as in step 2 above. Fill the container with water from a volumetric measure of the same volume as the labeled volume. Measure the distance to the liquid level at the center of the container and record this level in Column 7 below the reading recorded in step 2. If this distance is equal to or greater than the distance determined in step 2, assume that the package is satisfactory. If the distance is less than the distance determined in step 2, the product may be short measure. When the audit test indicates that short measure is possible use the "Violation Procedure" in Section 3.7.1.

(Amended 2010)

3.7.1. Violation Procedure

a. **How is it determined if the containers meet the package requirements?**

Use the following method if the liquid level is within the measuring range of the micrometer.

Note: Do not shake or invert the containers selected as the sample.

Steps:

1. The first step is to follow Section 2.3.1. "Define the Inspection Lot" to determine which "Category A" sampling plan to use; select a random sample; and then use the following procedure. The steps noted with an (*) are required if there is paint adhering to the lid and it cannot be removed by scraping into the can.

2. Determine the gross weight of these packages and record in Column 2 of the "Example Worksheet for Possible Violation in Checking Paint" worksheet. (in this section).

3. Record the labeled volume of the first tare sample package in Column 1 of the worksheet. Use a circular metal disk to eliminate can "sag" and remove the lid. If paint clings to the lid of the container, scrape it off with a spatula

4.* If paint that adheres to the lid cannot be completely removed by scraping the paint into the can, determine the weight of the lid plus any adhering paint. Clean the paint lid with solvent and weigh again. Subtract the clean lid weight from the lid weight with paint to determine the weight of the paint adhering to the lid. Record this weight in Column 3.

5. Place the spanning bar and depth gage across the top of the paint can. Mark the location of the spanning bar on the rim of the paint container. Measure the distance to the liquid level at the center of the container to the nearest 20 μm (0.02 mm) (0.001 in). Record the distance in Column 4.

6. Empty and clean the sample container and lid with solvent; dry and weigh the container and lid. Record the tare weight in Column 5.

7. Set up the container in the same manner as in step 2.

8. Place the spanning bar at the same location on the rim of the paint container as marked in step 5. With the depth gage set as described in step 5, deliver water into the container in known amounts until the water reaches the same level occupied by the paint as indicated by the depth gage. Record this volume of water (in mL or fl oz) in Column 6 of the worksheet. This is the volume occupied by the paint in the container. Follow steps, 9a, 10a, and 11a if scraping does not remove the paint from the lid. In order to determine if gravimetric testing can be used to test the other packages in the sample, follow only steps 9, 10, and 11 when no paint adheres to the lid.

9. Subtract the weight of the container (Column 5) from the gross weight (Column 2) to arrive at the net weight of paint in the selected container. Record the net weight in Column 7 of the worksheet.

Steps:

9a.* Subtract the weight of the container (Column 5) and the weight of product on the lid (Column 3) from the gross weight (Column 2) to arrive at the net weight of paint in the container. Record in Column 7 (excluding the weight of the paint on the lid).

10. Calculate the weight of the labeled volume of paint (for the first package opened for tare).

net weight (Column 7) x labeled volume (Column 1) ÷ volume of paint in can (Column 6)

Record this value in Column 8.

10a.* Calculate the package volume =

*volume in can (Column 6) + (lid paint weight [Column 3] x
volume in can [Column 6] ÷ net weight [Column 7])*

Record it in Column 9 of the worksheet.

11. Calculate the package error. Use the following formula if paint does not adhere to the lid.

Package error = (Column 6 value) − (labeled volume)

11a.* Use the following formula if paint does adhere to the lid and will not come off by scraping.

Package error = (Column 9 value) − (labeled volume)

12. Repeat steps 2 through 11 for the second package chosen for tare.

Example Worksheet for Possible Violation in Checking Paint (add additional rows as needed)								
1. Labeled Volume	2. Gross Weight	3. Lid Paint Weight (Wet − Dry)	4. Liquid Level	5. Tare	6. Water Volume	7. Net Wt. = 2 − 5	8. Weight of Labeled Volume = 7 x 1 ÷ 6	9. Package Volume = 6 + [(3 ÷ 7) x 6)]

b. **When can a gravimetric procedure be used?**

A gravimetric procedure is used if the weights of the labeled volume for the first two packages do not differ from each other by more than one division on the scale (if they meet this criterion, check the rest of the sample gravimetrically and record in Column 8).

c. **How is "nominal gross weight" determined?**

Use Section 2.3.6. to determine the "Nominal Gross Weight" as follows:

The nominal gross weight equals the sum of the average weight of the labeled volume (average of values recorded in Column 8) plus the average tare (average of values recorded in Column 3) for the packages selected for tare. Note that the weight of a given volume of paint often varies considerably from container to container; therefore, volumetric measurements may prove necessary for the entire sample.

Evaluation of Results

Follow the procedures in Section 2.3.7. "Evaluating Results" to determine lot conformance.

3.8. Testing Viscous Materials – Such As Caulking Compounds and Pastes

a. **How are viscous materials such as caulking compounds and paste tested?**

Use the following procedure for any package of viscous material labeled by volume. It is suitable for very viscous materials such as cartridge-packed caulking compounds, glues, pastes, and other similar products. It is best to conduct this procedure in a laboratory using a hood to ventilate solvent fumes. If used in the field, use in a well ventilated area. Except for the special measurement procedures to determine the weight of the labeled volume, this procedure follows the basic test procedure.

Steps:

1. For each weight of a known volume determination, pack a portion of the packaged product into a pre-weighed cup of known volume (called a "density cup" or "pycnometer") and weigh.

2. From the weight of the known volume, determine the weight of the labeled volume.

3. Compare the nominal gross weight with the gross weight to determine the package error.

b. **What type of measurement equipment is needed to test packages of caulk, pastes, and glues?**

- A scale that meets the requirements in Section 2.2. "Measurement Standards and Test Equipment."

- Pycnometer, a vessel of known volume used for weighing semifluids. The pycnometer can be bought or made. If it is made, refer to it as a "density cup." To make a 150 mL or 5 fl oz density cup, cut off the lip of a 150 mL beaker with an abrasive saw and grind the lip flat on a lap wheel. The slicker plate is available commercially. The metrology laboratory should calibrate the density cup gravimetrically with respect to the contained volume using the procedure in ASTM E542-01 (2007), "Standard Practice for Calibration of Laboratory Volumetric Apparatus."

- Appropriate solvents (water, Stoddard solvent, kerosene, alcohol, etc.)

- Caulking gun (for cartridge packed products)

c. **How is a pycnometer prepared for use?**

Before using, the metrology laboratory should calibrate the pycnometer (or the density cup and slicker plate) with respect to volume (mL or fl oz). If applicable, comply with any special instructions furnished by the manufacturer to calibrate a pycnometer that has not been calibrated. It is not necessary to reweigh or recalibrate for each test; however, mark the pieces of each unit to prevent interchange of cups and slicker plates.

d. **How is it determined if the containers meet the package requirements?**

Steps:

Follow the procedures in Section 2.3.1. "Define the Inspection Lot." Use a "Category A" sampling plan in the inspection; select a random sample; then, use the following procedure to determine lot compliance.

1. Weigh a calibrated pycnometer and slicker plate and record as "pycnometer weight" and record the volume of the pycnometer.

2. Determine the gross weight of the first package and record the weight value. Open the package and transfer the product to the pycnometer by filling it to excess. Use a caulking gun to transfer product from the caulking cartridges. If using a pycnometer, cover it with a lid and screw the cap down tightly. Excess material will be forced out through the hole in the lid, so the lid must be clean. If using a density cup, place the slicker plate over ¾ of the cup mouth, press down and slowly move the plate across the remainder of the opening. With the slicker plate in place, clean all the exterior surfaces with solvent and dry.

3. Completely remove the product from the package container; clean the package container with solvent; dry and weigh it to determine the tare weight.

4. Weigh the filled pycnometer or filled density cup with slicker plate and record this weight. Subtract the weight of the empty pycnometer from the filled weight to determine the net weight of the product contained in the pycnometer and record this weight.

5. Clean the pycnometer and repeat steps 3, 4, and 5 for the second package in the tare sample.

   Determine acceptability of the density variation on the two packages selected for tare. If the difference between the densities of both packages exceeds one division of the scale, do not use the gravimetric procedure to determine the net quantity of contents. Instead, use the procedure in steps 8 and 9.

Note: If the gravimetric procedure can be used, perform steps 8 and 10.

Steps:

6. Calculate the weight of product corresponding to the labeled volume of product according to the following formula:

Product Density X Labeled Volume = Labeled Weight
Weight of Product in Pycnometer ÷ Pycnometer Volume = Product Density

7. Test each package individually by determining the product density in each package using the pycnometer and record the gross, tare, and net weight of each package. Subtract the weight of the labeled volume (determined for each package) from the net weight of product to arrive at each individual package error in units of weight.

8. Convert the package errors to units of volume using the following formula:

Package Error (volume) =
(Package Error [weight] x Pycnometer Volume) ÷ (Weight of Product in Pycnometer)

9. Record the package errors on the report form, using an appropriate unit of measure.

Evaluation of Results

Follow the procedures in Section 2.3.7. "Evaluating Results" to determine lot conformance.

3.9.   Peat Moss

a.   **How are packages of peat and peat moss labeled by compressed volume tested?**

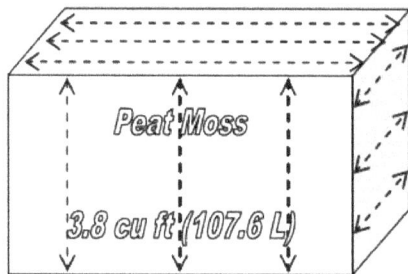

Figure 3-1. Peat Moss

Steps:

1. Measure the dimensions of the compressed material to determine if it contains the labeled quantity.

2. Take three measurements (both ends and middle) of each dimension and calculate their average.

3. Multiply the averages to obtain the compressed cubic volume.

4. For each dimension (length, width, and height) take three equidistant measurements, take the average of each respective dimension and multiply to determine the cubic measure as follows:

Average height X average width X average length = cubic measurement

5. Compare measured volume with labeled volume to determine package error.

(Amended 2010)

b. **How are packages of peat and peat moss labeled by uncompressed volume tested?**

Use the following method to test peat moss sold using an uncompressed volume as the declaration of content. The procedure is based on ASTM D2978-03, "Standard Method of Test for Volume of Processed Peat Materials."

Test Equipment

- 12.7 mm (or ½ in) sieve

- Use one of the following measures as appropriate for the package size. (Refer to Table 3-4. "Specifications for Test Measures for Mulch and Soils" for additional information on test measure construction.)

  - 28.3 L (1 ft³) measure with inside dimensions of 30.4 cm (12 in) by 30.4 cm (12 in) by 30.4 cm (12 in). Mark the inside of the measure with horizontal lines every 1.2 cm (½ in) so that package errors can be directly determined

  - 100 L (3.5 ft³) measure with inside dimensions of 50 cm (19.68 in) by 50 cm (19.68 in) by 40 cm (15.74 in). The inside of the measure should be marked with horizontal lines every 1.2 cm (½ in) so that package errors can be directly determined

- Straight edge, 50.8 cm (20 in) in length

- Sheet for catching overflow of material

- Level (at least 15.24 cm (6 in) in length)

Test Procedure

Steps:

1. Follow Section 2.3.1. "Define the Inspection Lot." Use a "Category A" sampling plan in the inspection; select a random sample; then, use the following procedure to determine lot compliance.

2. Open each package in turn, remove the contents, and pass them through the sieve directly into the measuring container (overfilling it). Use this method for particulate solids (such as soils or other garden materials) labeled in cubic dimensions or dry volume. Some materials may not pass through the sieve for peat moss; in these instances, separate the

Steps:

materials by hand (to compensate for packing and settling of the product after packaging) before filling the measure.

Note: Separated material (product not passing through the sieve) must be included in the product volume.

3. Shake the measuring container with a rotary motion at one rotation per second for 5 seconds. Do not lift the measuring container when rotating it. If the package contents are greater than the measuring container capacity, level the measuring container with a straightedge using a zigzag motion across the top of the container.

4. Empty the container. Repeat the filling operations as many times as necessary, noting the partial fill of the container for the last quantity delivered using the interior horizontal markings as a guide.

5. Record the total volume.

6. To compute each package error, subtract the labeled quantity from the total volume and record it.

Evaluation of Results

Follow the procedures in Section 2.3.7. "Evaluating Results" to determine lot conformance.

3.10. Mulch and Soils Labeled by Volume

a. **What products are defined as mulch and soil?**

- Mulch is defined as "any product or material except peat or peat moss that is advertised, offered for sale, or sold for primary use as a horticultural, above-ground dressing, for decoration, moisture control, weed control, erosion control, temperature control, or other similar purposes."

- Soil is defined as "any product or material, except peat or peat moss that is advertised or offered for sale, or sold for primary use as a horticultural growing media, soil amendment, and/or soil replacement."

b. **What type of measurement equipment is needed to test packages of mulch and soil?**

- A test measure appropriate for the package size that meets the specifications for test measures in Table 3-4. "Specifications for Test Measures for Mulch and Soils"

- Drop cloth/polyethylene sheeting for catching overflow of material

- Level (at least 15 cm [6 in] in length)

Table 3-4. Specifications for Test Measures for Mulch and Soils						
Nominal Capacity of Test Measure[4]	Actual Volume of the Measure[4]	Interior Wall Dimensions[1]			Marked Intervals on Interior Wall[3]	Volume Equivalent of Marked Intervals
		Length	Width	Height[2]		
30.2 L (1.07 cu ft) for testing packages that contain less than 28.3 L (1 cu ft or 25.7 dry qt)	31.9 L (1.13 cu ft)	213.4 mm (8.4 in)	203.2mm (8 in)	736.6 mm (29 in)	12.7 mm (½ in)	524.3 mL (32 in³)
28.3 L (1 cu ft)	28.3 L (1 cu ft)	304.8 mm (12 in)	304.8 mm (12 in)	304.8 mm (12 in)		1179.8 mL (72 in³)
56.6 L (2 cu ft)	63.7 L (2.25 cu ft)	304.8 mm (12 in)	304.8 mm (12 in)	685.8 mm (27 in)		
		406.4 mm (16 in)	228.6 mm (9 in)	685.8 mm (27 in)		
84.9 L (3 cu ft)	92 L (3.25 cu ft)	304.8 mm (12 in)	304.8 mm (12 in)	990.6 mm (39 in)		
		406.4 mm (16 in)	228.6 mm (9 in)	990.6 mm (39 in)		

Measures are typically constructed of 1.27 cm (½ in) marine plywood. A transparent sidewall is useful for determining the level of fill, but must be reinforced if it is not thick enough to resist distortion. If the measure has a clear front, place the level gage at the back (inside) of the measure so that the markings are read over the top of the mulch.

Notes
[1] Other interior dimensions are acceptable if the test measure approximates the configuration of the package under test and does not exceed a base configuration of the package cross-section.

[2] The height of the test measure may be reduced, but this will limit the volume of the package that can be tested.

[3] When lines are marked in boxes, they should extend to all four sides of the measure if possible to improve readability. It is recommended that a line indicating the MAV level also be marked to reduce the possibility of reading errors when the level of the mulch is at or near the MAV.

[4] The Nominal Capacity is given to identify the size of packages that can be tested in a single measurement using the dry measure with the listed dimensions. It is based on the most common package sizes of mulch in the marketplace. If the measures are built to the dimensions shown above the actual volume will be larger than the nominal volume so that plus errors (overfill) can be measured accurately.

(Amended 2010)

c. How is it determined if the packages meet the package requirements?

Use the following procedure:

Steps:

1. Follow the Section 2.3.1. "Define the Inspection Lot." Use a "Category A" sampling plan in the inspection, and select a random sample, then use the following procedure to determine lot conformance.

2. Open each package in turn. Empty the contents of the package into a test measure and level the contents by hand. Do not rock, shake, drop, rotate, or tamp the test measure. Read the horizontal marks to determine package net volume.

Note: Some types of mulch are susceptible to clumping and compacting. Take steps to ensure that the material is loose and free flowing when placed into the test measure. Gently roll the bag before opening to reduce the clumping and compaction of material.

3. Exercise care in leveling the surface of the mulch/soil and determine the volume reading from a position that minimizes errors caused by parallax.

d. How are package errors determined?

Determine package errors by subtracting the labeled volume from the package net volume in the measure. Record each package error.

$$Package\ Error = Package\ Net\ Volume - Labeled\ Volume$$

Evaluation of Results

Follow the procedures in Section 2.3.7. "Evaluating Results" to determine lot conformance.

Note: In accordance with Appendix A, Table 2-10. Exceptions to the Maximum Allowable Variations for Textiles, Polyethylene Sheeting and Film, Mulch and Soil Labeled by Volume, Packaged Firewood and Packages Labeled by Count with 50 Items or Fewer, apply an MAV of 5 % of the declared quantity to mulch and soil sold by volume. When testing mulch and soil with a net quantity in terms of volume, one package out of every 12 in the sample may exceed the 5 % MAV (e.g., one in a sample of 12 packages; two in a sample of 24 packages; four in a sample of 48 packages). However, the sample must meet the average requirement of the "Category A" Sampling Plan.

3.11. Ice Cream Novelties

Note: The following procedure can be used to test packaged products that are solid or semisolid and that will not dissolve in, mix with, absorb, or be absorbed by the fluid into which the product will be immersed. For example, ice cream labeled by volume can be tested using ice water or kerosene as the immersion fluid.

Exception: Pelletized ice cream is beads of ice cream which are quick frozen with liquid nitrogen. The beads are relatively small, but can vary in shape and size. On April 17, 2009, the FDA issued a letter stating that this product is considered semisolid food, in accordance with 21 CFR 101.105(a). The FDA also addresses that the appropriate net quantity of content declaration for pelletized ice cream products be in terms of net weight.

(Added 2010)

a. **How are ice cream novelties inspected to see if the labeled volume meets the package requirements?**

Use the following volume displacement procedure that uses a displacement vessel specifically designed for ice cream novelties such as ice cream bars, ice cream sandwiches, or cones. The procedure determines the volume of the novelty by measuring the amount of water displaced when the novelty is submerged in the vessel. Two displacements per sample are required to subtract the volume of sticks or cups.

The procedure first determines if the densities of the novelties are the same from package to package (in the same lot) so that a gravimetric test can be used to verify the labeled volume. If a gravimetric procedure is used, compute an average weight for the declared volume from the first two packages and weigh the remainder of the sample. If the gravimetric procedure cannot be used, use the volume displacement procedure for all of the packages in the sample.

Test Equipment

- A scale that meets the requirements in Section 2.2. "Measurement Standards and Test Equipment"

- Volumetric measures

- Displacement vessel with dimensions that is appropriate for the size of novelties being tested. Figure 3-2, *Example of a Displacement Vessel* shows an example of a displacement vessel. It includes an interior baffle that reduces wave action when the novelty is inserted and the downward angle of the overflow spout reduces dripping. Other designs may be used.

Figure 3-2. Example of a Displacement Vessel

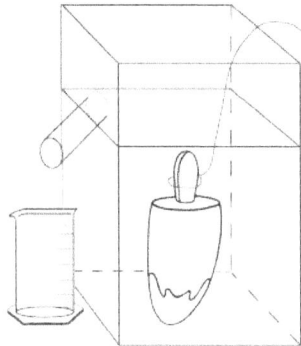

Note: This displacement vessel can be constructed or similar devices may be obtained from any laboratory equipment or science education suppliers. The U.S. Department of Commerce does not endorse or recommend any particular device over similar commercially available products from other manufacturers.

- Thin wire, clamp, or tongs

- Freezer or ice chest and dry ice

- Single-edged razor or sharp knife (for sandwiches only)

- Ice water/kerosene maintained at 1 °C (33 °F) or below

- Indelible marker (for ice pops only)

- Level, at least 15.24 cm (6 in) in length

- A partial immersion thermometer (or equivalent) with a range of − 1 °C to + 50 °C (30 °F to 120 °F), at least 1 °C (1 °F) graduations, and with a tolerance of ± 1 °C (± 2 °F)

- A table-top, laboratory-type jack of sufficient size to hold the displacement vessel

- Stopwatch

Test Procedure

Steps:

1. Follow the procedures in Section 2.3.1. "Define the Inspection Lot." Use a "Category A" sampling plan in the inspection; select a random sample; then use the following steps to determine lot compliance.

2. Maintain the samples at the reference temperature for frozen products that is specified in Table 3-1. "Reference Temperatures for Liquids" (i.e., − 18 °C [0 °F]). Place the samples in the freezer or ice chest until they are ready to be tested, and then remove packages from the freezer one at a time.

3. According to the type of novelty, prepare the sample products as follows:

    ➤ **Ice-pop**. Mark on the stick(s) with the indelible marker the point to which the pop will be submerged in the ice water. (After the ice-pop contents have been submerged, remove the novelty to determine the volume of the stick.)

    ➤ **Cone**. Make a small hole in the cone below the ice cream portion to allow air to escape.

    ➤ **Sandwich**. Determine whether the declared volume is (a) the total volume of

Steps:

> the novelty (that is, including the cookie portion) or (b) the volume of the ice-cream-like portion only. If the declared volume is the volume of only the ice-cream-like portion, shave off the cookie with a razor or knife, leaving some remnants of cookie to ensure that no ice cream is accidentally shaved off. Work quickly, and return the novelty to the freezer before the sandwich softens.

> ➤ **Cup**. Remove the cap from the cup. (After the cup and novelty contents have been submerged, remove the novelty from the cup to determine the volume of the cup.)

b. **How is it determined if the ice cream novelty packages meet the requirements in this handbook?**

Steps:

1. Follow Section 2.3.1. "Define the Inspection Lot." Use a "Category A" sampling plan in the inspection; select a random sample; then use the following procedure to determine lot compliance.

2. Fill the displacement vessel with ice water until it overflows the spout. Allow it to sit until dripping stops. Raise the displacement vessel as necessary and place the graduate beneath the spout.

3. Remove a package from the freezer, determine its gross weight and record it.

4. Submerge the novelty as suggested until it is below the surface level of the water.

> ➤ **Ice-pop**. Use a clamp, tongs, or your fingers to hold the stick(s) and submerge the pop to the level marked in step 3 of the Test Procedures.

> ➤ **Cone**. Shape the wire into a loop, and use it to push the cone, headfirst (ice cream portion first) into the ice water. Do not completely submerge the cone immediately: let water fill the cone through the hole made in step 3 of the Test Procedures before completely submerging the novelty.

> ➤ **Sandwich or cup**. Skewer the novelty with the thin wire or form a loop on the end of the wire to push the sandwich or ice-cream portion or cup completely below the liquid level.

5. Record the total water volume in the graduate. For a cone or sandwich, record the water volume as the net volume and go to step 8. For ice-pops or cups, record the water volume in the graduate as the gross volume and go to step 7.

6. Refill the displacement vessel with water to overflowing and reposition the empty graduate under the spout.

> ➤ **Ice-pop**. Melt the ice pop off the stick or sticks. Submerge the stick or sticks to the line marked in step 5. Record the volume of tare material (i.e., stick) by measuring the water displaced into the graduate. The net volume for the ice-pop

Steps:

is the gross volume recorded in step 6 minus the volume of the tare materials in this step. Record this volume as the "volume of novelty." To determine the error in the package, subtract the labeled quantity from the volume of novelty.

> **Cup**. Remove the novelty from the cup. Rinse the cup, and then submerge it in the displacement vessel. Small pinholes in the base of the cup can be made to make submersion easier. Record the volume of water displaced into the graduate by the cup as the volume of tare material. The net volume for the novelty is the gross volume determined in step 6 minus the volume of the tare materials determined in this step. Record this as the net volume of the novelty. To determine the error in the package, subtract the labeled quantity from the volume of novelty.

7. Clean and air-dry the tare materials (sticks, wrappers, cup, lid, etc.). Weigh and record the weight of these materials for the package.

8. Subtract the tare weight from the gross weight to obtain the net weight and record this value.

9. Compute the weight of the labeled volume for the package using the following formula and then record the weight:

   Product Density = (gross weight in item 4) ÷ (the total water volume in step 6)
   Weight of labeled volume = (labeled volume) x (Product Density)

10. Repeat steps 4 through 10 for a second package.

11. If the weight of the labeled volume in steps 11 and step 12 differ from each other by more than one division on the scale, the gravimetric test procedure cannot be used to test the sample for compliance. If this is the case, steps 3 through 7 for each of the remaining packages in the sample must be used to determine their net volumes and package errors. Then go to evaluation of results.

c. **How is "nominal gross weight" determined?**

Steps:

1. Use Section 2.3.5. "Tare Procedure" to determine the Average Used Dry tare Weight of the sample.

2. Using the weight of labeled volume determined in step 10 calculate the Average Product Weight by adding the densities of the product from the two packages and dividing the sum by two.

   Average Product Density Weight = Labeled Volume x Average Product Density

3. Calculate the "nominal gross weight" using the formula:

   Nominal Gross Weight = Average Product Weight + Average Used Dry Tare Weight

d. **How are the errors in the sample determined**?

Steps:

1. Weigh the remaining packages in the sample.

2. Subtract the nominal gross weight from the gross weight of each package to obtain package errors in terms of weight.

Note: Compare the sample packages to the nominal gross weight.

3. Follow Section 2.3. "Basic Test Procedure."

To convert the average error or package error from weight to volume, use the following formula:

Package Error in Volume = (Package Error in Weight) ÷ (Product Density)

Evaluation of Results

Follow the procedures in Section 2.3.7. "Evaluating Results" to determine lot conformance.

3.12. Fresh Oysters Labeled by Volume

a. **What requirements apply to packages of fresh oysters labeled by volume**?

Packaged fresh oysters removed from the shell must be labeled by volume. The maximum amount of permitted free liquid is limited to 15 % by weight. Testing the quantity of contents of fresh oysters requires the inspector to determine total volume, total weight of solids and liquid, and the weight of the free liquid.

Test Equipment

- A scale that meets the requirements in Section 2.2. "Measurement Standards and Test Equipment"

- Volumetric measures

- Micrometer depth gage (ends of rods fully rounded), 0 mm to 228 mm (0 in to 9 in)

- Strainer for determining the amount of drained liquid from shucked oysters. Use as a strainer a flat bottom metal pan or tray constructed to the following specifications:

  ➢ Sides: 5.08 cm (2 in)

  ➢ Area: 1935 cm^2 (300 in^2) or more for each 3.78 L (1 gal) of oysters

Note: Strainers of smaller area dimensions are permitted to facilitate testing smaller containers.

> ➢ Perforations:
>   Diameter: 6.35 mm (¼ in)
>   Location: 3.17 cm (1¼ in) apart in a square pattern, or perforations of equivalent area and distribution.

- Spanning bar, 2.54 cm by 2.54 cm by 30.48 cm (1 in by 1 in by 12 in)

- Rubber spatula

- Level, at least 15.24 cm (6 in) in length

- Stopwatch

b. **How is it determined if the containers meet the package requirements?**

Steps:

1. Follow the Section 2.3.1. "Define the Inspection Lot." Use a "Category A" sampling plan in the inspection; select a random sample; then, use the following test procedure to determine lot compliance.

2. Determine and record the gross weight of a sample package.

3. Set the container on a level surface and open it. Use a depth gage to determine the level of fill. Lock the depth gauge. Mark the location of the gauge on the package.

4. Weigh a dry 20.32 cm or 30.48 cm (8 in or 12 in) receiving pan and record the weight. Set strainer over the receiving pan.

5. Pour the contents from the container onto the strainer without shaking it. Tip the strainer slightly and let it drain for 2 minutes. Remove strainer with oysters. It is normal for oysters to include mucous (which is part of the product) that will not pass through the strainer, so do not force it.

6. Weigh the receiving pan and liquid and record the weight. Subtract the weight of the dry receiving pan from the weight of pan and liquid to obtain the weight of free liquid and record the value.

7. Clean, dry, and weigh the container and record the tare weight. Subtract the tare weight from the gross weight to obtain the total weight of the oysters and liquid and record this value.

8. Determine and record the percent of free liquid by weight as follows:

Percent of free liquid by weight = [(weight of free liquid) ÷ (weight of oysters + liquid)] X 100.

Steps:

9. Set up the depth gauge on the dry package container as in step 3. Pour water from the flasks and graduate as needed to re-establish the level of fill obtained in step 3. Add the volumes delivered as the actual net volume for the container and record the value.

Note: Some containers will hold the declared volume only when filled to the brim; they may have been designed for other products, rather than for oysters. If the net volume is short measure (per step 9), determine if the container will reach the declared volume only if filled to the brim. Under such circumstance, the package net volumes will all be short measure because the container cannot be filled to the brim with a solid and liquid mixture. A small headspace is required in order to get the lid into the container without losing any liquid.

Evaluation of Results

Follow the procedures in Section 2.3.7. "Evaluating Results" to determine lot conformance.

3.13. Determining the Net Contents of Compressed Gas in Cylinders

a. What type of compressed gases may be tested with these procedures?

These procedures are for industrial compressed gas. Compressed gas may be labeled by weight (for example, Liquefied Petroleum [LP] gas, or carbon dioxide) or by volume. Acetylene, liquid oxygen, nitrogen, nitrous oxide, and argon are all filled by weight. Acetylene is sold by liters or by cubic feet. Helium, gaseous oxygen, nitrogen, air, and argon are filled according to pressure and temperature tables.

b. What type of test procedures must be used?

Checking the net contents of compressed gas cylinders depends on the method of labeling; those labeled by weight are generally checked by weight. Cylinders filled by using pressure and temperature charts must be tested using a pressure gauge that is connected to the cylinder. Determine the volume using the pressure and temperature of the cylinder.

c. Should any specific safety procedures be followed?

Yes, be aware of the hazards of the high pressure found in cylinders of compressed gas. An inspector should handle compressed gas only if the inspector has been trained and is knowledgeable regarding the product, cylinder, fittings, and proper procedures (see *Compressed Gas Association [CGA] pamphlet P-1, "Safe Handling of Compressed Gases in Containers,"* for additional information). Additional precautions that are necessary for personal safety are described in the CGA Handbook of Compressed Gases. All personnel testing compressed gases should have this manual for reference and be familiar with its contents. It is essential that the inspector be certain of the contents before connecting to the cylinder. Discharging a gas or cryogenic liquid through a system for which the material is not intended could result in a fire and/or explosion or property damage due to the incompatibility of the system and the product. Before connecting a cylinder to anything, be certain of the following:

1. Always wear safety glasses.

2. The cylinder is clearly marked or labeled with the correct name of the contents and that no conflicting marks or labels are present. Do not rely on the color of the cylinder to identify the contents of a cylinder. Be extremely careful with all gases because some react violently when mixed or when coming in contact with other substances. For example, oxygen reacts violently when it comes in contact with hydrocarbons.

3. The cylinder is provided with the correct Compressed Gas Association (CGA) connection(s) for the product. A proper connection will go together smoothly; so excessive force should not be used. Do not use an adapter to connect oxygen to non-oxygen cleaned equipment. When a cylinder valve is opened to measure the internal pressure, position the body away from the pressure gauge blowout plug or in front of the gauge if the gauge has a solid cast front case. If the bourdon tube should rupture, do not be in a position to suffer serious injuries from gas pressure or fragments of metal.

Note: The acetone in acetylene cylinders is included in the tare weight of the cylinder. Therefore, as acetylene is withdrawn from the cylinder, some acetone will also be withdrawn, changing the tare weight.

4. Thoroughly know the procedure and place emphasis on safety precautions before attempting any tests. Do not use charts referred to in the procedure until the necessary training has been completed. When moving a cylinder, always place the protective cap on the cylinder. Do not leave spaces between cylinders when moving them. This can lead to a "domino" effect if one cylinder is pushed over.

5. Open all valves slowly. A failure of the gauge or other ancillary equipment can result in injuries to nearby persons. Remember that high gas pressure can propel objects with great force. Gas ejected under pressure can also cause serious bodily injuries if someone is too close during release of pressure.

6. One of the gauges will be reserved for testing oxygen only and will be prominently labeled "For Oxygen Use Only." This gauge must be cleaned for oxygen service and maintained in that "clean" condition. The other gauge(s) may be used for testing a variety of gases if they are compatible with one another.

7. Observe special precautions with flammable gas in cylinders in addition to the several precautions necessary for the safe handling of any compressed gas in cylinders. Do not "crack" cylinder valves of flammable gas before connecting them to a regulator or test gauge. This is extremely important for hydrogen or acetylene.

d. **What type of measurement equipment is needed to test cylinders of compressed gas?**

Test Equipment

- Use a scale that meets the requirements in Section 2.2. "Measurement Standards and Test Equipment." Use a wooden or non-sparking metal ramp to roll the cylinders on the scale to reduce shock loading.

- Two calibrated precision bourdon tube gauges or any other approved laboratory-type pressure-measuring device that can be accurately read within plus or minus 40 kPa (5 psi). A gauge having scale increments of 200 kPa (25 psi) or smaller shall be considered as satisfactory for reading within plus or minus 40 kPa (5 psi). The range of both gauges shall be a minimum of 0 kPa to 23 MPa (0 psi to 5000 psi) when testing cylinders using standard industrial cylinder valve connections. These standardized connections are listed in *"CGA Standard V-1, Standard for Compressed Gas Cylinder Valve Outlet and Inlet Connections for use with Gas Pressures up to 21 MPa (3000 psi)."* For testing cylinders with cylinder valve connections rated for over 21 MPa (3000 psi), the test gauge and its inlet connection must be rated at 14 MPa (2000 psi) over the maximum pressure that the connection is rated for in CGA V-1.

Notes:

1. There are standard high-pressure industrial connections on the market that are being used up to their maximum pressure of 52 MPa (7500 psi).

2. Any gauge or connectors used with oxygen cylinders must be cleaned for oxygen service, transported in a manner which will keep them clean and never used for any other gas including air or oxygen mixtures. Oxygen will react with hydrocarbons and many foreign materials that may cause a fire or explosion.

- An approved and calibrated electronic temperature measuring device or three calibrated liquid-in-glass thermometers having either a digital readout or scale division of no more than 1 °F (0.5 °C). The electronic device equipped with a surface temperature sensor is preferred over a liquid-in-glass thermometer because of its shorter response time.

- Two box-end wrenches of 29 mm ($1\frac{1}{8}$ in) for oxygen, nitrogen, carbon dioxide, argon, helium, and hydrogen and 22 mm ($\frac{7}{8}$ in) for some sizes of propane. All industrial CGA connections are limited to these two hex sizes. Avoid using an adjustable wrench because of the tendency to round the edges of the fittings, which can lead to connections not being tightened properly.

- Use a separate gauge and fitting for each gas to be tested. If adapters must be used, do not use on oxygen systems.

3.13.1. Test Procedure for Cylinders Labeled by Weight

a. **How is it determined if the containers meet the package requirements using the gravimetric test procedure?**

Steps:

1. Follow Section 2.3.1. "Define the Inspection Lot." Use a "Category A" sampling plan in the inspection; select a random sample; then use the following test procedure to determine lot compliance.

2. The cylinder should be marked or stenciled with a tare weight. The marked value may or may not be used by the filling plant when determining the net weight of those cylinders sold or filled by weight. If there is a tare weight marked on the net contents tag or directly on the cylinder, then an actual tare weight was determined at the time of

Steps:

fill. If there is no tare weight marked on a tag or on the cylinder, then the stamped or stenciled tare weight is presumed to have been used to determine the net contents.

Note: Check the accuracy of the stamped tare weights on empty cylinders whenever possible. The actual tare weight must be within (a) ½ % of the stamped tare weight for 9.07 kg (20 lb) tare weights or less or (b) ¼ % of the stamped tare weight for greater than 9.07 kg (20 lb) tare weights. (See NIST Handbook 130, "Method of Sale Regulation.")

3. Place cylinder on scale and remove protective cap. The cap is not included in the tare weight. Weigh the cylinder and determine net weight, using either the stamped or stenciled tare weight, or the tare weight marked on the tag. Compare actual net weight with labeled net weight, or use the actual net weight to look up the correct volume declaration (for Acetylene Gas), and compare that with the labeled volume.

Most producers will replace acetone in the cylinder before the cylinder is refilled, filling the cylinder with acetone to the stamped tare weight. Other producers, although not following recommended procedures, do not replace the acetone until it drops to a predetermined weight. In the latter situation, the refilling plant must note the actual tare weight of the cylinder and show it on the tag containing the net content statement or on the cylinder itself. Refer to tables for acetylene if necessary (if the acetylene is labeled by volume).

3.13.2 Test Procedure for Cylinders Labeled by Volume

a. **How is it determined if the containers meet the package requirements using the volumetric test procedure?**

Steps:

1. Follow Section 2.3.1. "Define the Inspection Lot." Use a "Category A" sampling plan in the inspection; select a random sample; then use the following test procedure to determine lot compliance.

2. Determine the temperature of the cylinders in the sample. Place the thermometer approximately halfway up a cylinder in contact with the outside surface. Take the temperature of three cylinders selected at random and use the average temperature of the three values.

3. Using the appropriate pressure gauge, measure the pressure of each cylinder in the sample.

4. Determine the cylinder nominal capacity from cylinder data tables or from the manufacturer. (These tables must be obtained in advance of testing.)

5. Using NIST Technical Note 1079 "Tables of Industrial Gas Container Contents and Density for Oxygen, Argon, Nitrogen, Helium, and Hydrogen" (available on-line at (http://www.nist.gov/pml/wmd), determine the value (SCF/CF) from the content tables at

Steps:

the temperature and pressure of the cylinder under test.

6. Multiply the cylinder nominal capacity by the value (SCF/CF) obtained from the content tables. This is the actual net quantity of gas.

7. Subtract the labeled net quantity from the actual net quantity to determine the error.

Evaluation of Results

Follow Section 2.3.7. "Evaluating Results" to determine lot conformance.

3.14. Firewood

3.14.1. Volumetric Test Procedure for Packaged Firewood with a Labeled Volume of 113 L (4 ft³) or Less

a. **How are packages of firewood tested?**

Follow Section 2.3.1. "Define the Inspection Lot." Use a "Category A" sampling plan in the inspection; select a random sample, then use the test procedure provided in Section 3.14.3. "Crosshatched Firewood" to determine lot compliance.

Unless otherwise indicated, take all measurements without rearranging the wood or removing it from the package. If the layers of wood are crosshatched or not ranked in discrete sections in the package, remove the wood from the package, re-stack, and measure accordingly.

Test Equipment

- Linear Measure. Take all measurements in increments of 0.5 cm (³/₁₆ in) or less and round up.

- Binding Straps. Binding straps are used to hold wood bundles together if the bundles need to be removed from the package/wrapping material.

3.14.2. Boxed Firewood

a. **How is the volume of firewood contained in a box determined?**

Steps:

1. Follow Section 2.3.1. "Define the Inspection Lot." Use a "Category A" sampling plan in the inspection; select a random sample; then use the following test procedure to determine lot conformance.

2. Open the box to determine the average height of wood within the box; measure the

Steps:

internal height of the box. Take three measurements (record as "d_1, d_2. . .etc.") along each end of the stack. Measure from the bottom of a straightedge placed across the top of the box to the highest point on the two outermost top pieces of wood and the center-most top piece of wood. Round measurements down to the nearest 0.5 cm ($\frac{1}{8}$ in). If pieces are obviously missing from the top layer of wood, take additional height measurements at the highest point of the uppermost pieces of wood located at the midpoints between the three measurements on each end of the stack. Calculate the average height of the stack by averaging these measurements and subtracting from the internal height of the box according to the following formula.

Average Height of Stack =
(Internal Height of Box) − (sum of measurements) ÷ (number of measurements)

3. Determine the average width of the stack of wood in the box by taking measurements at three places along the top of the stack. Measure the inside distance from one side of the box to the other on both ends and in the middle of the box. Calculate the average width.

Average Width = (W_1 + W_2 + W_3) ÷ (3)

4. To determine the average length of the pieces of wood, remove the wood from the box and select the five pieces with the greatest girth. Measure the length of each of the five pieces from center-to-center. Calculate the average length of the five pieces.

Average Length = (L_1 + L_2 + L_3 + L_4 + L_5) ÷ (5)

5. Calculate the volume of the wood within the box. Use dimensions for height, width, and length.

Volume in liters = (height in cm x width in cm x length in cm) ÷ (1000)

Volume in cubic feet = (height in inches x width in inches x length in inches) ÷ (1728)

6. For boxes of wood that are packed with the wood ranked in two discrete sections perpendicular to each other, calculate the volume of wood in the box as follows: (1) determine the average height, width, and length as in 1, 2 and 3 above for each discrete section, compute total volume, and (2) total the calculated volumes of the two sections. Take the width measurement for Volume 2 (V_2) from the inside edge of the box adjacent to V_2 to the plane separating VR_1 and V_2. Compute total volume by adding Volume 1 (V_1) and V_2 according to the following formula.

Total Volume = V_1 + V_2

7. Follow Section 2.3.7. "Evaluating Results" to determine lot conformance.

3.14.3. Crosshatched Firewood

a. **How must the volume of stacked or crosshatched firewood be measured?**

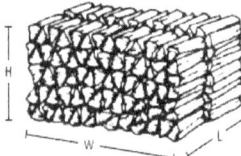

Figure 3-3. Stacked Firewood

Steps:

1. Follow Section 2.3.1. "Define the Inspection Lot." Use a "Category A" sampling plan in the inspection; select a random sample; and use the following test procedure to determine lot compliance.

2. Stack the firewood in a ranked and well-stowed geometrical shape that facilitates volume calculations (i.e., rectangular). The number of measurements for each dimension given below is the minimum that should be taken.

3. Determine the average measurements of the stack:

   ➢ Height: Start at one end of the stack; measure the height of the stack on both sides at four equal intervals. Calculate and record the average height.

   ➢ Length: Start at the base of the stack; Measure the length of the stack in four equal intervals. Calculate and record the average length.

   ➢ Width: Select the five pieces with the greatest girth. Measure the length of the pieces, calculate and record the average piece length.

4. Calculate Volume:

   Volume in liters = (Avg. Height [cm] × Avg. Width [cm] × Avg. Length in [cm]) ÷ 1000

   Volume in cubic feet = (Avg. Height [in] × Avg. Width [in] × Avg. Length [in]) ÷ 1728

5. Follow Section 2.3.7. "Evaluating Results" to determine lot conformance.

3.14.4. Bundles and Bags of Firewood

a. **How is the volume of bundles and bags of firewood measured?**

Figure 3-4. Bundle of Firewood

Steps:

1. Follow Section 2.3.1. "Define the Inspection Lot." Use a "Category A" sampling plan in the inspection; select a random sample; then use the following test procedure to determine lot compliance.

2. Average area of ends: secure a strap around each end of the bundle or bag of wood to prevent movement during testing and to provide a definite perimeter. Use two or more straps to secure the wood.

3. Set one end of the bundle or bag on tracing paper large enough to cover the end completely. Draw a line around the perimeter of the bundle or bag on the tracing paper.

4. Transfer the tracing paper to a template graduated in square centimeters or square inches. Count the number of square centimeters or square inches that are enclosed within the perimeter line. Estimate portions of square centimeters or square inches not completely within the perimeter line to the nearest one-quarter square inch.

5. Repeat this process on the opposite end of the bundle or bag.

6. Calculate the Average Area:

$$Average\ Area = (Area\ 1 + Area\ 2) \div 2$$

7. Average length of the pieces of wood – select the five pieces with the greatest girth and measure the length of the pieces. Calculate the average length of the pieces of wood:

$$Average\ Length = (L_1 + L_2 + L_3 + L_4 + L_5) \div 5$$

8. Calculate Volume:

$$Volume\ in\ liters = (Average\ Area\ [cm^2] \times Average\ Length\ [cm]) \div 1000$$

$$Volume\ in\ cubic\ feet = (Average\ Area\ [in^2] \times Average\ Length\ [in]) \div 1728$$

Evaluation of Results

Follow Section 2.3.7. "Evaluating Results" to determine lot conformance.

Note: Specified in Appendix A, Table 2-10. "Exceptions to the Maximum Allowable Variations for Textiles, Polyethylene Sheeting and Film, Mulch and Soil Labeled by Volume, Packaged Firewood, and Packages Labeled by Count with 50 Items or Fewer." – Maximum allowable variations for individual packages are not applied to packages of firewood.

THIS PAGE INTENTIONALLY LEFT BLANK

Chapter 4. Test Procedures – Packages Labeled by Count, Linear Measure, Area, Thickness, and Combinations of Quantities

4.1.   Scope

a.   What types of packaged goods can be tested using these procedures?

Use these procedures to determine the net contents of products sold by count, area, thickness, and linear measure. If a package includes more than one declaration of quantity, each declaration must meet the package requirements.

b.   Can the gravimetric test procedure be used to verify the net quantity of contents of packages labeled by count and linear measure?

Use the gravimetric procedure (below) to test products sold by measure or count if the density of the product does not vary excessively from one package to another.

c.   What procedures may be used if the gravimetric test procedure cannot be used?

Open each package in the sample and measure or count the items.

4.2.   Packages Labeled by Count

a.   How are packages labeled by count tested?

If the labeled count is 50 items or fewer, use Section 4.3. "Packages Labeled with 50 Items or Fewer." If the labeled count is more than 50 items, see Section 4.4. "Packages Labeled by Count of More than 50 Items." If the labeled count is more than 50 items for corn, soybeans, field beans, and wheat seeds, see Section 4.11 "Procedure for Checking the Contents of Specific Agricultural Seed Packages Labeled by Count."
(Amended 2010)

b.   Can a gravimetric test procedure be used to verify the labeled count of a package?

Yes, if the scale being used is sensitive enough to determine the weight of individual items. Use the following procedures to determine if the sample packages can be tested gravimetrically.

Steps:
1.   For packages labeled with a count of 84 or higher, calculate the weight equivalent for the MAV/6 for the labeled count of the package. MAV/6 must be at least equal to one-half

77

Steps:

scale division on a mechanical scale or one division on a digital scale.

2. For packages with a labeled count of 83 or fewer, when each unit weighs at least 2 scale divisions, consider the scale acceptable.

> **Example:** *According to Appendix A, Table 2-7. Maximum Allowable Variations (MAVs) for Packages Labeled by Count, the MAV is 7 for a package labeled with a count of 250 items. The scale should be capable of measuring differences corresponding to MAV/6 or, in this example, the weight of one item.*

> ➤ If the scale meets the appropriate requirement, gravimetric testing can be used to determine package count or,

> ➤ If the scale does not meet the criteria, count the content in each package in the sample.

4.3. Packages Labeled with 50 Items or Fewer

Test Procedure

Steps:

1. Follow Section 2.3.1. "Define the Inspection Lot." Use a "Category A" sampling plan in the inspection; select a random sample; then use the following test procedure to determine lot compliance.

2. Open the packages and count the number of items in each. Record the number of packages that contain fewer than the labeled count.

Evaluation of Results

1. For the sample size indicated in Column 1 of Appendix A, Table 2-11. "Accuracy Requirements for Packages Labeled by Low Count of (50 or Fewer) and Packages Given Tolerance (Glass and Stemware)," refer to Column 2 to determine the number of packages that are allowed to contain fewer than the labeled count.

2. If the number of packages in the sample that contain fewer than the labeled count exceeds the number permitted in Column 2, the sample and the lot fail to meet the package requirement.

Note: For statistical reasons, the average requirement does not apply to packages labeled by count of 50 or fewer items, **and the MAV does not apply to the lot**. It only applies to the packages in the sample.

3. Maximum Allowable Variations: The MAVs listed in Appendix A, Table 2-7. "Maximum Allowable Variations (MAVs) for Packages Labeled by Count" define the limits of reasonable variation for an individual package even though the MAV is not directly used in the sampling plan. Individual packages that are undercount by more than the MAV are considered defective. Even if the sample passes, these should be repacked, relabeled, or otherwise handled.

Example: *If testing a lot of 160 packages of pencils labeled "50 pencils," choose a random sample of 12 packages from the lot. If the scale cannot discriminate between differences in count, open every package and count the pencils. For example, assume the 12 package counts are: 50, 52, 50, 50, 51, 53, 52, 50, 50, 50, 47, and 50.*

Because only one package contains fewer than 50 pencils, the sample passes the test (refer to Appendix A, Table 2-11, "Accuracy Requirements for Packages Labeled by Low Count [50 or Fewer] and Packages Given Tolerances [Glass and Stemware]"). However, the package containing 47 pencils should not be introduced into commerce even though the lot complies with the package requirements because it is undercount by more than the MAV (1 item) permitted in Appendix A, Table 2-7, "Maximum Allowable Variations (MAVs) for Packages Labeled by Count."

4.4. Packages Labeled by Count of More than 50 Items

Test Procedures

There are two procedures to determine count without opening all packages in the sample. Both use the weight of a counted number of items in the package. If the weight of discrete items or numbers of items in a package varies excessively, the packaged items must be counted rather than weighed.

Test Equipment

Use a scale that meets the requirements in Section 2.2. "Measurement Standards and Test Equipment."

4.4.1. Audit Procedure

Use this procedure to audit lots of packages labeled by count of more than 50 items. Determine the lot compliance based on actual count or Section 4.4.2. Procedures to "Use if the Inspector Suspects the Lot Violates the Package Requirements."

Note: The precision of this procedure is only ± 1 %.

Steps:

1. Follow Section 2.3.1. "Define the Inspection Lot." Use a "Category A" sampling plan in the inspection; select a random sample; then use the following test procedure to determine lot compliance.

2. Select an initial tare sample according to Section 2.3.5. "Tare Procedures."

3. Gross weigh the first package in the tare sample and record this weight.

4. Select the number of items from the first tare package that weighs the greater:

   ➢ 10 % of the labeled count; or

Steps:

> ➤ a quantity equal to at least 50 minimum divisions on the scale.

> **Example:** *Using a scale with 1 g divisions, the selected count must weigh at least 50 grams. If a scale with 0.001 lb divisions is used, the selected count must weigh at least 0.05 lb. Record the count and weight.*

5. Calculate the weight of the labeled count using the following formula:

$$\textit{Weight of the Labeled Count} =$$
$$\textit{(labeled count} \times \textit{weight of items in step 4)} \div \textit{(Count of items in step 4)}$$

Record the result as "labeled count weight."

6. Gross weigh the remaining packages of the tare sample and keep contents of opened packages separated in case all of the items must be counted.

7. Determine the Average Used Dry Tare Weight of the sample according to Section 2.3.5. "Tare Procedures."

8. The weight of the labeled count plus the average tare weight represents the "nominal gross weight."

9. Subtract the nominal gross weight from the gross weight of the individual packages and record the errors.

$$\textit{(Package error [weight])} =$$
$$\textit{(actual package gross weight)} - \textit{(nominal gross weight)}$$

10. Convert the package errors in units of weight to count:

Package error (count) = (Package error [weight] x labeled count) ÷ (labeled count weight)

Round any fractional counts up to whole items in favor of the packager. Record the package error in units of count. Compute the average error.

> ➤ If the average error is minus, go to the "procedure to use if the inspector suspects the lot violates the package requirements" below.

> ➤ If the average error is zero or positive, the sample is presumed to conform to the package requirements.

4.4.2. Procedures to Use if the Inspector Suspects the Lot Violates the Package Requirements

If possible, use the gravimetric procedure to determine compliance. To minimize the number of packages to be opened, combine the measurement of the weight of the number of units in the package with the determination of tare. Therefore, it will not be necessary to open more packages than the tare sample. If the audit procedure in this section has been used, the possible violation procedure below can be followed with the same sample if package contents have been kept separate and can still be counted. Use the following steps to determine if the sample passes or fails.

Steps:

1. Follow Section 2.3.1. "Define the Inspection Lot." Use a "Category A" sampling plan in the inspection; select a random sample; then use the following test procedure to determine lot compliance. Use a scale that meets the criteria specified in 4.2. "Packages Labeled by Count."

2. Select an initial tare sample according to Section 2.3.5. "Tare Procedures."

3. Gross weigh the packages selected for the tare sample and record these weights. Open these packages and determine the tare and net weights of the contents, and count the exact number of items in the packages. Record this information.

4. Calculate and record the weights of the labeled counts for the first two packages using the formula:

   Weight of labeled count = (labeled count) × (contents weight ÷ contents count)

   To avoid round off errors, carry at least two extra decimal places in the calculation until the weight of the labeled count is obtained. To use the gravimetric procedure, the difference in weights of the labeled counts of the two packages must not exceed one scale division.

   ➢ If the difference in weights exceeds this criterion, determine the actual count per package for every package in the sample recording plus and minus errors. Then, follow the procedures in Section 2.3.7. "Evaluating Results" to determine lot conformance.

   ➢ If the difference is within the criterion, average the weights of the labeled count and go on to step 5.

5. Determine the Average Used Dry Tare Weight of the sample according to provisions in Section 2.3.5. "Tare Procedures."

6. Determine and record the nominal gross weight by adding the average weight of the labeled count of items in the package step 4 to the average tare weight step 5.

7. Weigh the remaining packages in the sample, subtract the nominal gross weight from the gross weight of the individual packages, and record the errors.

   Package Error (weight) = (Actual Package Gross Weight) − (Nominal Gross Weight)

Steps:

8. Look up the MAV for the package size from Appendix A, Table 2-7. "Maximum Allowable Variations (MAVs) for Packages Labeled by Count" and convert it to weight using the formula:

MAV (weight) =
(MAV (count) × Average Weight of Labeled Count [from step 4]) ÷ (Labeled Count)

Convert the MAV to dimensionless units by dividing the MAV (weight) by the unit of measure and record.

Evaluation of Results

Follow the procedures in Section 2.3.7. "Evaluation Results" to determine lot conformance.

Convert back to count when completing the report form, using the following formula:

Average Package Error (count) = (Avg. Pkg. Error [dimensionless units]) × (Unit of Measure) ×
(Labeled Count) ÷ (Average. Weight of Labeled Count)

4.5. Paper Plates and Sanitary Paper Products

a. How are the labeled dimensions of paper plates and sanitary paper products verified?

Follow Section 2.3.1. "Define the Inspection Lot." Use a "Category A" sampling plan in the inspection; select a random sample; then use the following procedure to determine lot compliance.

The following procedures are used to verify the size of paper plates and other products. The following procedure may be used to verify the size declarations of other disposable dinnerware.

Note: Do not distort the item's shape during measurement.

The count of sanitary paper products cannot be adequately determined by weighing. Variability in sheet weight and core weight requires that official tests be conducted by actual count. However, weighing can be a useful audit method. These products often declare total area as well as unit count and sheet size. If the actual sheet size measurements and the actual count comply with the average requirements, the total area declaration is assumed correct.

Test Equipment

- Steel tapes and rules. Determine measurements of length to the nearest division of the appropriate tape or rule.

  - Metric Units:

    For labeled dimensions 40 cm or less, linear measure: 30 cm in length, 1 mm divisions; or a 1 m rule with 0.1 mm divisions, overall length tolerance of 0.4 mm.

For labeled dimensions greater than 40 cm, 30 m tape with 1 mm divisions.

> ➤ Inch-pound Units:

For labeled dimensions 25 in or less, use a 36 in rule with $^1/_{64}$ in or $^1/_{100}$ in divisions and an overall length tolerance of $^1/_{64}$ in.

For dimensions greater than 25 in, use a 100 ft tape with in divisions and an overall length tolerance of 0.1 in.

- Measuring Base

Note: A measuring base may be made of any flat, sturdy material approximately 38 cm (15 in) square. Two vertical side pieces approximately 3 cm (1 in) high and the same length as the sides of the measuring base are attached along two adjoining edges of the measuring base to form a 90° corner. Trim all white borders from two or more sheets of graph paper (10 divisions per centimeter or 20 divisions per inch). Place one sheet on the measuring base and position it so that one corner of graph paper is snug in the corner of the measuring base and vertical sides. Tape the sheet to the measuring base. Overlap other sheets on the first sheet so that the lines of top and bottom sheet coincide, expanding the graph area to a size bigger than plates to be measured; tape these sheets to the measuring base. Number each line from the top and left side of base plates: 1, 2, 3, etc.

b. **How are paper products inspected?**

Steps:
1. Follow Section 2.3.1. "Define the Inspection Lot." Use a "Category A" sampling plan in the inspection; select a random sample; then use the following test procedure to determine lot compliance.

2. Select an initial tare sample according to Section 2.3.5. "Tare Procedure."

3. Open each package and select one item from each.

Note: Some packages of plates contain a combination of different-sized plates. In this instance, take a plate of each declared size from the package to represent all the plates of that size in the package. For example, if three sizes are declared, select three different plates from each package.

c. **How are paper products measured?**

Note: Occasionally, packages of plates declared to be one size contain plates that can be seen by inspection to be of different sizes in the same package. In this instance, select the smallest plate and use the methods below to determine the package error. If the smallest plate is not short measure by more than the MAV, measure each size of plate in the package and calculate the average dimensions.

> **Example:** *If 5 plates measure 21.41 cm (8.43 in) and 15 measure 21.74 cm (8.56 in), the average dimension for this package of 20 plates is 21.66 cm (8.53 in).*

Steps:

1. For paper plates: Place each item on the measuring base plate (or use the linear measure) with the eating surface down so two sides of the plate touch the sides of the measuring base. For other products, use either the measuring base or a linear measure to determine actual labeled dimensions (e.g., packages of napkins, rolls of paper towels). If testing folded products, be sure that the folds are pressed flat so that the measurement is accurate.

2. If the measurements reveal that the dimensions of the individual items vary, select at least 10 items from each package. Measure and average these dimensions. Use the average dimensions to determine package error in step 3 below.

3. The package error equals the actual dimensions minus the labeled dimensions.

Evaluation of Results

Follow the procedures in Section 2.3.7. "Evaluating Results" to determine lot conformance.

4.6. Special Test Requirements for Packages Labeled by Linear or Square Measure (Area)

a. Are there special measurement requirements for packages labeled by dimensions?

Yes, products labeled by length (such as yarn) or area, often requires the application of tension to the ends of the product in order to straighten the product before measuring. When testing yarn and thread, apply tension and use the specialized equipment specified in ASTM D1907-07, "Standard Test Method for Linear Density of Yarn (Yarn Number) by the Skein Method," in conjunction with the sampling plans and package requirements described in this handbook.

Evaluation of Results

Follow the procedures in Section 2.3.7. "Evaluating Results" to determine lot conformance.

4.7. Polyethylene Sheeting

a. Which procedures are used to verify the declarations on polyethylene sheeting and bags?

Follow Section 2.3.1. "Define the Inspection Lot." Use a "Category A" sampling plan in the inspection; select a random sample; then use the following test procedure to determine lot compliance.

Note: Most polyethylene products are sold by length, width, thickness, area, and net weight.

Test Equipment

*   A scale that meets the requirements in Section 2.2. "Measurement Standards and Test Equipment."

- Steel tapes and rules. Determine measurements of length to the nearest division of the appropriate tape or rule.

  ➢ Metric Units:

    For labeled dimensions 40 cm or less, linear measure: 30 cm in length, 1 mm divisions; or a 1 m rule with 0.1 mm divisions, overall length tolerance of 0.4 mm.

    For labeled dimensions greater than 40 cm, 30 m tape with 1 mm divisions.

  ➢ Inch-pound Units:

    For labeled dimensions 25 in or less, use a 36 in rule with $^1/_{64}$ in or $^1/_{100}$ in divisions and an overall length tolerance of $^1/_{64}$ in.

    For dimensions greater than 25 in, use a 100 ft tape with $^1/_{16}$ in divisions and an overall length tolerance of 0.1 in.

- Deadweight dial micrometer (or equal) equipped with a flat anvil, 6.35 mm or ($^1/_4$ in) diameter or larger, and a 4.75 mm ($^3/_{16}$ in) diameter flat surface on the head of the spindle. The anvil and spindle head surfaces should be ground and lapped, parallel to within 0.002 mm (0.0001 in), and should move on an axis perpendicular to their surfaces. The dial spindle should be vertical, and the dial should be at least 50.8 mm (2 in) in diameter. The dial indicator should be continuously graduated to read directly to 0.002 mm (0.0001 in) and should be capable of making more than one revolution. It must be equipped with a separate indicator to indicate the number of complete revolutions. The dial indicator mechanism should be fully jeweled. The frame should be of sufficient rigidity that a load of 1.36 kg (3 lb) applied to the dial housing, exclusive of the weight or spindle presser foot, will not cause a change in indication on the dial of more than 0.02 mm (0.001 in). The indicator reading must be repeatable to 0.001 2 mm (0.000 05 in) at zero. The mass of the probe head (total of anvil, weight 102 g or [3.6 oz], spindle, etc.) must be 113.4 g (4 oz). The micrometer should be operated in an atmosphere free from drafts and fluctuating temperature and should be stabilized at ambient room temperature before use.

- Gage blocks covering the range of thicknesses to be tested should be used to check the accuracy of the micrometer

- T-square

Test Procedure

  Steps:
  1. Follow Section 2.3.1. "Define the Inspection Lot." Use a "Category A" sampling plan in the inspection; select a random sample; then use the following test procedure to determine lot compliance.

Steps:

2. Be sure the product is not mislabeled. Check the label declaration to confirm that all of the declared dimensions are consistent with the required standards. The declaration on sheeting, film, and bags shall be equal to or greater than the weight calculated by using the formulas below. Calculate the final value to four digits and declare to three digits dropping the final digit (e.g., if the calculated value is 2.078 lb, then the declared net weight is truncated to 2.07 lb).

Example Label:

┌───┐
│ **Polyethylene Sheeting**   │
│   │
│ **1.82 m (6 ft) x 30.48 m (100 ft)**   │
│   │
│ **101.6 μm (4 mil)**   │
│   │
│ **5.03 kg (11.1 lb)**   │
└───┘

3. Use the following formulas to compute a target net weight. The labeled weight should equal or exceed the target net weight or the package is not in compliance.

   ➤ For metric dimensions:

   $$Target\ Mass\ in\ Kilograms = (T \times A \times D) \div 1\ 000$$

   Where: T = *nominal thickness in centimeters*

   A = *nominal length in centimeters* × *nominal width (the nominal width for bags is twice the labeled width) in centimeters*

   D = *density in grams per cubic centimeter* *

   ➤ For inch-pound dimensions:

   $$Target\ Weight\ in\ Pounds = T \times A \times D \times 0.036\ 13$$

   Where: T = *nominal thickness in inches;*

   A = *nominal area; that is the nominal length in inches* × *nominal width (the nominal width for bags is twice the labeled width) in inches;*

   D = *density in grams per cubic centimeter; 0.036 13 is a factor for converting* g/cm^3 *to* lb/in^3.

*Determined by ASTM Standard D1505-03, "Standard Method of Test for Density of Plastics by the Density Gradient Technique." For the purpose of this handbook, the minimum density shall be 0.92 g/cm^3 when the actual density is not known.

Evaluation

Steps:

1. Perform the calculations as shown in the following samples. If the product complies with the label declaration, go to step 2.

    Example: *Sample Calculations*

    ➢ *For metric units:*

    $$(0.010\ 16\ cm \times [(1.82\ m \times 100\ ^{cm}/_m) \times (30.48\ m \times 100\ ^{cm}/_m)] \times 0.92\ ^g/_{cm^3}) \div 1000\ ^g/_{kg}$$
    $$= a\ target\ weight\ of\ 5.18\ kg$$

    In this example, the labeled net mass of 5.03 kg does not meet the target net mass, so the product is not in compliance.

    ➢ *For inch-pound units:*

    $$(0.004\ in) \times [(6\ ft \times 12\ ^{in}/_{ft}) \times (100\ ft \times 12\ ^{in}/_{ft})] \times 0.92\ ^g/_{cm^3} \times 0.03613$$
    $$= a\ target\ weight\ of\ 11.48\ lb$$

    In this example, the labeled net weight of 11.1 lb does not meet the target net weight, so the product is not in compliance.

2. Select packages for tare samples. Determine and record the gross weights of the initial tare sample.

3. Extend the product in the sample packages to their full dimensions and remove by hand all creases and folds.

4. Measure the length and width of the product to the closest 3 mm ($^1/_8$ in). Make all measurements at intervals uniformly distributed along the length and width of the sample and record the results. Compute the average length and width, and record.

    ➢ With rolls of product, measure the length of the roll at three points along the width of each roll and measure the width at a minimum of 10 points along the length of each roll.

    ➢ For folded products, such as drop cloths or tarpaulins, make three length measurements along the width of the sample and three width measurements along the length of the sample.

5. Determine and record the average tare weight according to Section 2.3.5. "Tare Procedures."

4.7.1. Evaluation of Results – Length, Width, and Net Weight

Note: If the sample meets the package requirements for the declarations of length, width, and weight proceed to step 3 to verify the thickness declaration.

Steps:

1. Follow the procedures in Section 2.3.7. "Evaluating Results" to determine the lot conformance requirements for length, width, and weight.

2. If the sample failed to meet the package requirements for any of these declarations, no further measurements are necessary. The lot fails to conform.

3. Measure the thickness of the plastic sheet with a micrometer using the following guide. Place the micrometer on a solid level surface. If the dial does not read zero with nothing between the anvil and the spindle head, set it at zero. Raise and lower the spindle head or probe several times; it should indicate zero each time. If it does not, find and correct the cause before proceeding.

4. Take measurements at five uniformly distributed locations across the width at each end and five locations along each side of each roll in the sample. If this is not possible, take measurements at five uniformly distributed locations across the width product for each package in the sample.

5. When measuring the thickness, place the sample between the micrometer surfaces and lower the spindle head or probe near, but outside, the area where the measurement will be made. Raise the spindle head or probe a distance of 0.008 mm to 0.01 mm (0.000 3 in to 0.000 4 in) and move the sheet to the measurement position. Drop the spindle head onto the test area of the sheet.

6. Read the dial thickness two seconds or more after the drop, or when the dial hand or digital readout becomes stationary. This procedure minimizes small errors that may occur when the spindle head or probe is lowered slowly onto the test area.

7. For succeeding measurements, raise the spindle head 0.008 mm to 0.01 mm (0.000 3 in to 0.000 4 in) above the rest position on the test surface, move to the next measurement location, and drop the spindle head onto the test area. Do not raise the spindle head more than 0.01 mm (0.000 4 in) above its rest position on the test area. Take measurements at least 6 mm (¼ in) or more from the edge of the sheet.

8. Repeat steps 2 through 7 above on the remaining packages in the sample and record all thickness measurements. Compute and record the average thickness for the individual package and apply the following MAV requirements.

4.7.2. Evaluation of Results – Individual Thickness

Note: Refer to Appendix A, Table 2-10. Exceptions to the MAVs for Textiles, Polyethylene Sheeting
and Film, Mulch and Soil Labeled by Volume, Packaged Firewood, and Packages Labeled by Count with
50 Items or Fewer, and Specific Agricultural Seeds Labeled by Count.)
(Amended 2010)

- On polyethylene with a declared thickness greater than 25 μm (1 mil or 0.001 in): an individual
  thickness measured may be up to 20 % less than the declared thickness.

- No measured thickness of polyethylene labeled less than or equal to 25 μm (1 mil or 0.001 in),
  individual thickness measurements may be up to 35 % below the labeled thickness.

Count the number of values that are smaller than specified MAVs ($0.8 \times$ labeled thickness if 25 μm
[1 mil] or greater or $0.65 \times$ labeled thickness, if less than 25 μm [1 mil]). If the number of values that fail
to meet the thickness requirement exceeds the number of MAVs permitted for the sample size, the lot
fails to conform to requirements. No further testing of the lot is necessary. If the number of MAVs for
thickness measurements is less than or equal to the number permitted for the sample size, go on to
Evaluation of Results – Average Thickness.

4.7.3. Evaluation of Results – Average Thickness

The average thickness for any single package should be at least 96 % of the labeled thickness. This is an
MAV of 4 % (refer to Appendix A, Table 2-10. Exceptions to the MAVs for Textiles, Polyethylene
Sheeting and Film, Mulch and Soil Labeled by Volume, Packaged Firewood, and Packages Labeled by
Count with 50 Items or Fewer, and Specific Agricultural Seeds Labeled by Count.) This is an MAV of
4 %. Circle and count the number of package average thickness values that are smaller than
$0.96 \times$ labeled thickness. If the number of package average thicknesses circled exceeds the number of
MAVs permitted for the sample size, the lot fails to conform to requirements. No further testing of the lot
is necessary. If the number of MAVs for package average thickness is less than or equal to the number of
MAVs permitted for the sample size, proceed to Section 2.3.7. "Evaluating Results" to determine if the
lot meets the package requirements for average thickness.
(Amended 2010)

4.8. Packages Labeled by Linear or Square (Area) Measure

Test Equipment

- Use a scale or balance that meets the requirements in Section 2.2. "Measurement Standards and
  Test Equipment." Determine the suitability of the scale. Calculate the length or area of packaged
  product corresponding to MAV/6. If there is no suitable weighing device, all of the packages in
  the sample must be opened and measured.

- Steel tapes and rules – determine measurements of length to the nearest division of the
  appropriate tape or rule.

➢ Metric Units:

For labeled dimensions 40 cm or less, linear measure: 30 cm in length, 1 mm divisions; or a 1 m rule with 0.1 mm divisions, overall length tolerance of 0.4 mm.

For labeled dimensions greater than 40 cm, 30 m tape with 1 mm divisions.

➢ Inch-pound Units:

For labeled dimensions 25 in or less, use a 36 in rule with $\frac{1}{64}$ in or $\frac{1}{100}$ in divisions and an overall length tolerance of $\frac{1}{64}$ in.

For dimensions greater than 25 in, use a 100 ft tape with $\frac{1}{16}$ in divisions and an overall length tolerance of 0.1 in.

• T-square

Test Procedure

Steps:

1. Follow Section 2.3.1. "Define the Inspection Lot." Use a "Category A" sampling plan in the inspection; select a random sample; then use the following test procedure to determine lot compliance.

2. Select an initial tare sample according to Section 2.3.5. "Tare Procedures."

3. Gross weigh the first package in the tare sample and record this weight.

4. Determine and record the measurements (to the nearest division of the appropriate tape or rule) of the packaged goods (length, width, area; depending upon which dimensions are declared on the label) and weigh the goods from the first package opened for tare determination.

   ➢ Calculate and record the weight of the labeled measurements using the following formula:
   Weight of the labeled measurement =
   (labeled measurement) × (contents weight) ÷ (contents measurement)

   ➢ Look up and record the MAV in units of length or area measure (given in Appendix A, Table 2-8. "Maximum Allowable Variations for Packages Labeled by Length, (Width) or Area"

Note: See Appendix A, Table 2-10. "Exceptions to the MAVs for Textiles, and Polyethylene Sheeting and Film.

5. Determine and record the tare weight of the first package opened.

6. Determine and record the measurements (length, width, area; depending upon which dimensions are declared on the label) of the product in the second package chosen for

Steps:

tare determination (to the nearest division of the appropriate tape or rule). Determine and record the tare weight of this package.

7. Calculate and record the weight of the labeled measurement for the second package using the following formula:

Weight of the labeled measurement =
(labeled measurement) × (contents weight ÷ contents measurement)

The weights of the labeled measurement for two packages must not differ by more than one division on the scale. If they do, open all packages in the sample, measure individually, and compare them against the labeled measure to determine the package errors. If the criterion is met, go to step 8.

8. Calculate the average weight of the labeled measurement and record.

9. Determine and record the average tare weight according to Section 2.3.5, "Tare Procedures."

10. Compute and record the nominal gross weight by adding the average weight of the labeled measurements to the average tare weight.

11. Compute package errors according to the following formula:

Package error (weight) =
(actual package gross weight) − (nominal gross weight)

12. Convert the MAV to units of weight using the following formula:

MAV (weight) =
(Avg. Wt. of label measurements × MAV [length]) ÷ (labeled measurements)

Convert the MAV to dimensionless units by dividing the MAV (weight) by the unit of measure and record.

Evaluation of Results

Follow the procedure in Section 2.3.7, "Evaluating Results" to determine lot conformance.

Convert back to dimensions when completing the report form using the following the formula:

Average Package Error (dimension) = (Avg. Pkg. Error [dimensionless units]) × (Unit of Measure) ×
(Labeled unit of measure) ÷ (Avg. Weight of Labeled dimension)

4.9. Baler Twine – Test Procedure for Length

Test Equipment

- A scale that meets the requirements in Section 2.2. "Measurement Standards and Test Equipment," except a scale with 0.1 g (0.000 2 lb) increments must be used for weighing twine samples. The recommended minimum load for weighing samples is 20 divisions.

- Steel tapes and rules – Determine measurements of length to the nearest division of the appropriate tape or rule.

  ➢ Metric Units:

    For labeled dimensions 40 cm or less, linear measure: 30 cm in length, 1 mm divisions; or a 1 m rule with 0.1 mm divisions, overall length tolerance of 0.4 mm.

    For labeled dimensions greater than 40 cm, 30 m tape with 1 mm divisions.

  ➢ Inch-pound Units:

    For labeled dimensions 25 in or less, use a 36 in rule with $\frac{1}{64}$ in or $\frac{1}{100}$ in divisions and an overall length tolerance of $\frac{1}{64}$ in.

    For dimensions greater than 25 in, use a 100 ft tape with $\frac{1}{16}$ in divisions and an overall length tolerance of 0.1 in.

- A hand-held straight-face spring scale of at least 4.53 kg (10 lb) capacity or a cordage-testing device that applies the specified tension to the twine being measured. When measuring twine samples or total roll length, apply 4.53 kg (10 lb) of tension to the twine.

Test Procedure

Steps:

1. Follow Section 2.3.1. "Define the Inspection Lot." Use a "Category A" sampling plan in the inspection; select a random sample; then use the following test procedure to determine lot compliance.

2. Select packages for tare samples. Determine gross weights of the initial tare sample and record. Open the tare samples. Use the procedures for tare determination in Section 2.3.5. "Tare Procedures" to compute the average tare weight and record this value.

3. Procedure for obtaining twine samples: Randomly select four balls of twine from the packages that were opened for tare.

   From each of the four balls of twine:

   ➢ Measure and discard the first 10.05 m (33 ft) of twine from each roll. Accurate measurement requires applying tension to the ends of the twine before measuring

Steps:

in order to straighten the product.

➢ Take two 30.48 m (100 ft) lengths of twine from inside each roll.

➢ Weigh and record the weight of each piece separately and record the values.
Compare the weight values to determine the variability of the samples. If the
individual weights of the eight twine samples vary by more than one division on
the scale, use one of the following steps: If the lot is short, determine the actual
length of the lightest-weight roll found in the lightest-weight package of the lot
to confirm that the weight shortages reflect the shortages in the length of the
rolls; or, determine the average weight-per-unit of measure by taking ten
30.48 m (100 ft) lengths from inside the lightest weight package. Use this value
to recalculate its length and determine lot compliance.

4. Weigh all of the sample lengths together and record the total value. Determine the total
length of the samples (243.8 m or 800 ft, unless more than eight sample-lengths were
taken) and record the value. Compute the average weight-per-unit-of-length by dividing
the total weight by the total length of the pieces.

5. Determine the MAV for a package of twine (refer to Appendix A, Table 2-8. "Maximum
Allowable Variations for Packages Labeled by Length, Width, or Area").

➢ Record the total declared package length.

➢ Multiply the MAV from Appendix A, Table 2-8. "Maximum Allowable
Variations for Packages Labeled by Length, (Width), or Area," times the total
package length to obtain the MAV for length and record this value.

➢ Multiply the weight per unit of length (from step 3) times the MAV for the total
declared package length to obtain the MAV by weight and record this value.

➢ Convert the MAV to dimensionless units and record.

6. Calculate the nominal gross weight and record.

Follow Section 2.3.6. "Determine Nominal Gross Weight and Package Errors for Sample
Tare" to determine individual package errors. Determine errors using the following
formula:

Package error (weight) = (package gross weight) − (nominal gross weight)

➢ To convert the Package error in weight back to length, divide the weight by the
average weight-per-unit-of-length.

Evaluation of Results

Follow the procedures in Section 2.3.7. "Evaluating Results" to determine lot compliance.

4.10. Procedure for Checking the Area Measurement of Chamois

Chamois is natural leather made from skins of sheep and lambs that have been oil-tanned. Chamois are irregularly shaped, which makes area measurement difficult. Because of these characteristics, an accurate area determination can only be made using an internationally recognized method of conditioning (rehydrating) and measurement. Chamois is produced in a wet manufacturing process, so it has high moisture content at time of measurement. Chamois is hydroscopic; therefore, its dimensions and total area change as it loses or absorbs moisture. It is also subject to wrinkling. Because of the variation of the thickness and density, and therefore the weight per unit area of chamois, an estimated gross weight procedure cannot be used to verify the labeled area declaration.

Standard Test Conditions: As with all hydroscopic products, reasonable variations in measure must be allowed if caused by ordinary and customary exposure to atmospheric conditions that normally occur in good distribution practice. Both federal and international standards specify procedures to restore the moisture content of chamois so that tests to verify dimensions and area can be conducted.

Federal Test Method Standard 311, "Leather, Methods of Sampling and Testing," (January 15, 1969) defines the standard atmospheric condition for chamois as 50 ± 4 % relative humidity and 23 ± 2 °C (73.4 ± 3.6 °F). The chamois is considered to be at equilibrium moisture when the difference in two successive weighings, made at 1 hr intervals, is no greater than 0.25 % (e.g., the maximum change in weight on a 100 g sample in two successive weighings is less than 0.25 g (250 mg).

Test Procedures

The area of chamois is verified using a two-stage test procedure. The first stage is a field audit using the template test procedure. This test is used for field audits because it is simpler to perform and does not require the chamois to be conditioned. The field audit is used to identify chamois that are potentially under measure. It is not as accurate as the gravimetric procedure because some error results from reading the area from the template. The gravimetric procedure should be used for compliance testing because it includes conditioning (rehydrating) the chamois.

Template Test Method (for field audits)

Select a random sample of chamois and use the Template Procedure (below) to determine the area of each sample. Chamois is labeled in uniform sizes in terms of square decimeters and square feet, and are sized in increments of ¼ ft^2 (e.g., 1 ft^2, 1¼ ft^2, and 1½ ft^2). Separate the chamois into different sizes and define the inspection lot by specific sizes.

Test Equipment

Use a transparent, flexible template that is graduated in square centimeters or square inches and that has been verified for accuracy. The template must be large enough to completely cover the chamois under test.

Template Procedures

Steps:

1. Template Procedure

    Place the template over the chamois specimen on a smooth surface. Determine the area by counting the number of squares that cover the surface of the chamois. Estimate parts of the template that do not completely cover the chamois by adding the number of partially covered blocks. (See Figure 4-1.) Compute the total area and go to Evaluation to determine if further action is necessary.

First Stage – Decision Criteria

If the average minus error exceeds 3 % of the labeled area, the chamois may not be labeled accurately. To confirm the finding, the sample must be taken to a laboratory for conditioning and testing using the gravimetric test procedure.

2. Gravimetric Procedure for Area Measurement

    This test cannot be performed in the field because the samples must be conditioned with water before testing. This method is intended for use in checking full or cut skins, or pattern shapes. Open and condition all of the packages in the sample before determining their area on the recommended paper. Conditioning and verifying chamois can be accomplished without destroying the product. When successful tests are completed, the chamois may be repackaged for sale, so do not destroy the packaging material.

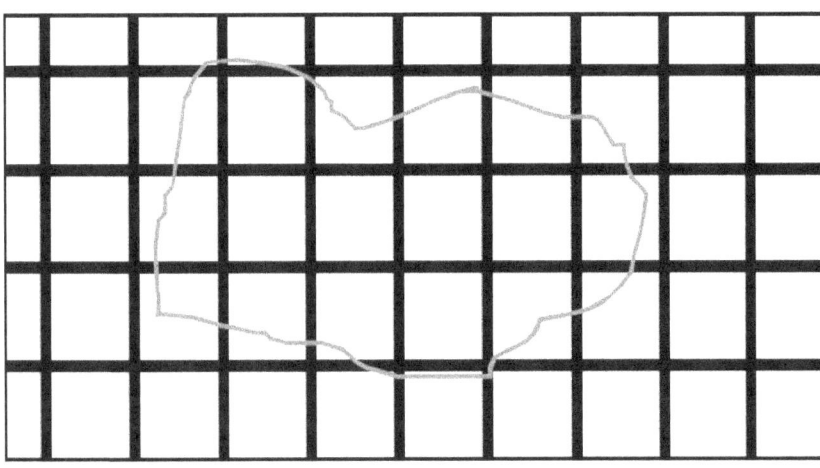

Figure 4-1. Template for Checking the Area of Chamois

Test Equipment

*   Scale with a capacity of 1 kg that is accurate to at least ± 0.01 g and a load-receiving element of adequate size to properly hold the chamois

*   Atomizer or trigger-type sprayer and sealable, airtight polyethylene bags

- Medium weight drawing paper (e.g., drawing paper, medium weight (100 lb), regular surface or comparable)

- Household iron with low temperature settings 30 °C to 40 °C (86 °F to 104 °F)

- Rule or tape that is graduated in centimeters or inches

- Instrument for cutting paper (razor blade, scissors, or cutting board)

Sample Conditioning

Steps:

1. Remove each sample from its package and weigh and record each weight. Using an atomizer-type sprayer, spray water in the amount of 25 % of the weight of each skin uniformly over its area. Place wetted chamois in an airtight polyethylene bag; seal the bag, and leave it in this condition at room temperature for 24 hours.

2. Open the bag, remove the chamois, and reweigh the chamois to confirm that it retained maximum moisture. (This is done by confirming that the difference in the two consecutive weighings conducted an hour apart does not exceed 0.25 %).

3. Place the chamois flat on a continuous piece of drawing paper. To remove wrinkles and make the chamois lie flat, use a normal domestic iron that is heated to a maximum of 30 °C to 40 °C (86 °F to 104 °F). Place the iron on the bottom of the skin, and iron the skin up from the center to the top. Then, iron the skin from the center out to each side. Iron until the skin is fully extended and perfectly flat.

Measurement

Steps:

1. Immediately after ironing, carefully draw around the outline of the skin on the paper. Remove the skin; carefully cut along the outline of the skin; weigh the cutout pattern, and record to the nearest 0.1 g as Sample Weight 1 (W1).

2. Lay out the pattern and cut an accurately measured rectangle of a size not less than one-half the area of the pattern. Weigh the cutout rectangle and record the weight to the nearest 0.1 g as Sample Weight 2 (W2). Calculate the area of the rectangle cut from the patterns by multiplying length by width and record as Area (A) in centimeters or square inches.

   ➢ For metric units – calculate the area of the original skin being checked as follows:

   $$W1/W2 \times A = Skin\ Area\ in\ cm^2/100 = Area\ in\ dm^2$$

Steps:

> For inch-pound units – calculate the area of the original skin being checked as
> follows:

$$W1/W2 \; x \; A = Skin \; Area \; in \; in^2/144 = Area \; ft^2$$

Evaluation of Results

Compute the average error for the sample and follow the procedures in Section 2.3.7, "Evaluating
Results" to determine lot conformance.

The MAV for area declarations on chamois is 3 % of the labeled area as specified in Appendix A,
Table 2-8, "Maximum Allowable Variations for Packages Labeled by Length, (Width), or Area".

4.11. Procedure for Checking the Contents of Specific Agriculture Seed Packages Labeled by Count

a. **How is the number of seeds determined in a sample of soybean, corn, wheat, and field bean, when using a mechanical seed counter?**

The following method shall be employed when using a mechanical seed counter to determine the
number of seeds contained in a sample of soybean (*Glycine max*), corn (*Zea mays*), wheat (*Triticum
aestivum*) and field bean (*Phaseolus vulgaris*).

Test Equipment

- Mechanical seed counter.

- Moisture proof container.

Test Procedure

Steps:

1. Testing samples shall be received and retained in moisture proof containers until the
   weight of the sample prepared for purity analysis is recorded. The sample shall be of at
   least 500 grams for soybean, corn, field beans, and 100 grams for wheat.

2. The seed counter shall be calibrated daily prior to use.

   > Prepare a calibration sample by counting 10 sets of 100 seeds. Visually examine each
   > set to insure that it contains whole seeds. Combine the 10 sets of seeds to make a 1000
   > seed calibration sample. The seeds of the calibration sample should be approximately
   > the same size and shape as the seeds in a sample being tested.

Note: If the seeds in a sample being tested are noticeably different in size or shape from those
in the calibration sample, prepare another calibration sample with seeds of the appropriate size
and shape. Periodically re-examine the calibration samples to insure that no seeds have been

Steps:

lost or added.

> ➤ Carefully pour the 1000 seed calibration sample into the seed counter. Start the counter and run it until all the seeds have been counted.

Note: The seeds should not touch as they run through the counter. Record the number of seeds as displayed on the counter read out.

> ➤ The seed count should not vary more than ± 2 seeds from 1000. If the count is not within this tolerance, clean the mirrors, adjust the feed rate and/or reading sensitivity. Rerun the calibration sample until it is within the ± 2 seed tolerance.

Note: If the seed counter fails the calibration procedure and sample has been checked to ensure that it contains 1000 seeds, do not use the counter until it has been repaired.

3. Immediately after opening the container, mix and divide the sample to obtain a sample for purity analysis (refer to Appendix D: AOSA Rules for Testing Seeds).

4. Record the weight of this sample in grams to the appropriate number of decimal places.

5. Conduct the purity analysis to obtain pure seed for the seed count test.

6. After the seed counter has been calibrated, test the pure seed portion from the purity test and record the number of seeds in the sample.

7. Calculation of results.

> ➤ Calculate the number of seeds per pound to the nearest whole number using the following formula:

Number of seeds per pound = 453.6 g/lb × no. of seeds counted divided by the weight (g) of sample analyzed for purity

> ➤ Calculate the number of seeds per pound to the nearest whole number using the following formula:

Number of seeds per pound = 453.6 g/lb × no. of seeds counted divided by the weight (g) of sample analyzed for purity

8. Determine the Maximum Allowable Variation (MAV).

> ➤ Multiply the labeled seed count by 4 % for soybean, 2 % for corn, 5 % for field bean, and 3 % for wheat.

Note: Express the maximum allowable variation (the number of seeds) to the nearest whole number. Consider the results of two tests in accord with the maximum allowable variation if the difference, expressed as the number of seeds, is equal to or less than the maximum allowable variation.

Example:
Kind of seed: Corn
Label claim: 2275 seeds/lb.

Lab Test: Purity working weight = 500.3 g
Seed count of pure seed = 2479 seeds

Number of seeds per pound = 453.6 g/lb × 2479 seeds divided by 500.3 g = 2247.6 seeds/lb
    Rounded to the nearest whole number = 2248 seeds/lb

Calculate maximum allowable variation value for corn:
    multiply label claim by 2 %
    2275 seeds/lb × 0.02 = 45.5 seeds/lb;
    rounded to the nearest whole number = 46 seeds/lb

Determine the difference between label claim and lab test:
    2275 seeds/lb − 2248 seeds/lb = 27 seeds/lb

The difference between the lab test and the label claim is less than the maximum allowable variation (27 < 46); therefore, the two results are in accord with the maximum allowable variation.

(Added 2010)

THIS PAGE INTENTIONALLY LEFT BLANK

Appendix A. Tables

Table 1-1. Agencies Responsible for Package Regulations and Applicable Requirements			
Commodity	**Responsible Agency**	**NIST Handbook 133 Sampling Plans**	**Table of Maximum Allowable Variations**
Meat and Poultry	U.S. Department of Agriculture/Food Safety and Inspection Service and state and local weights and measures.	**Use Table 2-1.** Sampling Plans for Category A to test packages at other than point of pack. Use Table 2-2. Sampling Plans for Category B to test packages in federally inspected meat and poultry plants.	Table 2-9. U.S. Department of Agriculture, Meat and Poultry, Groups and Lower Limits for Individual Packages
Foods, drugs, and cosmetics subject to the Food, Drug, and Cosmetic Act including those packaged at the retail store level that have been in interstate commerce (e.g., seafood) or those made with ingredients that have been in interstate commerce	U.S. Food and Drug Administration and state and local weights and measures http://www.fda.gov	**Use Table 2-1.** Sampling Plans for Category A to test packages at all locations.	**Table 2-5.** MAVs for Packages Labeled by Weight **Table 2-6.** MAVs for Packages Labeled by Liquid or Dry Volume **Table 2-7.** MAVs for Packages Labeled by Count **Table 2-8.** MAVs for Packages Labeled by Length (Width) or Area **Table 2-10.** Exceptions to the MAVs for Textiles, Polyethylene Sheeting and Film, Mulch and Soil Labeled by Volume, Packaged Firewood, and Packages Labeled by
Food products not subject to the Federal Food, Drug, and Cosmetic Act, including meat and poultry products packaged at the retail store level	State and local weights and measures http://www.nist.gov/pml/wmd/		
Non-food Consumer Products	Federal Trade Commission http://www.ftc.gov		
Non-food Consumer and Non-consumer Products	State and local weights and measures		

Table 1-1. Agencies Responsible for Package Regulations and Applicable Requirements			
Commodity	**Responsible Agency**	**NIST Handbook 133 Sampling Plans**	**Table of Maximum Allowable Variations**
Alcohol and Tobacco Products	Alcohol and Tobacco Tax and Trade Bureau. State and local weights and measures http://www.ttb.gov		Count with 50 Items or Fewer, and Specific Agriculture Seeds Labeled by Count.
Pesticides	U.S. Environmental Protection Agency and state and local weights and measures http://www.epa.gov		

Table 2-1. Sampling Plans for Category A					
1	2	3	4	5	6
Inspection Lot Size	Sample Size	Sample Correction Factor	Number of Minus Package Errors Allowed to Exceed the MAV[1]	Initial Tare Sample Size[2]	
				Glass and Aerosol Packages	All Other Packages
1	1	Apply MAV	0[1]	2	2
2	2	8.985			
3	3	2.484			
4	4	1.591			
5	5	1.242			
6	6	1.049			
7	7	0.925			
8	8	0.836			
9	9	0.769			
10	10	0.715			
11	11	0.672			
12 to 250	12	0.635			
251 to 3 200	24	0.422		3	
More than 3 200	48	0.290	1[1]		
[1]For mulch and soils packaged by volume, see Table 2-10. Exceptions to the Maximum Allowable Variations – 1 package may exceed the MAV for every 12 packages in the sample. [2]If sample size is 11 or fewer, the initial tare sample size and the total tare sample size is 2 samples.					

Table 2-2. Sampling Plans for Category B for Use in USDA-Inspected Meat and Poultry Plants Only			
1	2	3	4
Inspection Lot Size	Sample Size	Initial Tare Sample Size	Number of Packages Allowed to Exceed the MAVs in Table 2-9
250 or Fewer	10	2	0
251 or More	30	5	

Table 2-3. Category A					
	Total Number of Packages in Tare Sample				
	Note: Total number of packages to be opened for tare determination Numbers include those packages opened for initial tare sample				
Sample Size	**12**	**24**		**48**	
Initial Tare Sample Size	**2**	**2**	**3**	**2**	**3**
Ratio of R_c/R_t					
If range of tare equals "zero," use Initial Tare Sample Size. If the ratio is "zero" based on a "zero" range of net weight, open all of the packages in the sample.	2	2	3	2	3
If the ratio is greater than 0 but less than or equal to 0.2	12	24	24	48	48
0.21 to 0.60	12	24	24	48	48
0.61 to 0.70	12	24	24	47	47
0.71 to 0.80	12	23	23	47	47
0.81 to 1.00	12	23	23	46	46
1.01 to 1.10	11	23	23	46	46
1.11 to 1.20	11	23	23	45	45
1.21 to 1.30	11	22	22	45	45
1.31 to 1.50	11	22	22	44	44
1.51 to 1.60	11	22	22	43	43
1.61 to 1.70	11	21	21	42	42
1.71 to 1.80	10	21	21	42	42
1.81 to 1.90	10	21	21	41	41
1.91 to 2.00	10	20	20	41	41
2.01 to 2.10	10	20	20	40	40
2.11 to 2.20	10	20	20	39	39
2.21 to 2.30	10	19	19	39	39
2.31 to 2.40	9	19	19	38	38
2.41 to 2.50	9	19	19	37	37
2.51 to 2.60	9	18	18	37	37
2.61 to 2.70	9	18	18	36	36
2.71 to 2.80	9	18	18	35	35
2.81 to 2.90	9	17	17	34	34

Table 2-3. Category A					
	Total Number of Packages in Tare Sample				
	Note: Total number of packages to be opened for tare determination Numbers include those packages opened for initial tare sample				
Sample Size	12	24		48	
Initial Tare Sample Size	2	2	3	2	3
Ratio of R_c/R_t					
2.91 to 3.00	8	17	17	34	34
3.01 to 3.10	8	17	17	33	33
3.11 to 3.30	8	16	16	32	32
3.31 to 3.40	8	16	16	31	31
3.41 to 3.50	8	15	15	30	30
3.51 to 3.60	7	15	15	30	30
3.61 to 3.70	7	15	15	29	29
3.71 to 3.90	7	14	14	28	28
3.91 to 4.00	7	14	14	27	27
4.01 to 4.10	7	13	13	27	27
4.11 to 4.20	7	13	13	26	26
4.21 to 4.30	6	13	13	25	25
4.31 to 4.40	6	12	12	25	25
4.41 to 4.60	6	12	12	24	24
4.61 to 4.70	6	12	12	23	23
4.71 to 4.80	6	11	11	23	23
4.81 to 4.90	6	11	11	22	22
4.91 to 5.00	5	11	11	22	22
5.01 to 5.10	5	11	11	21	21
5.01 to 5.10	5	11	11	21	21
5.11 to 5.20	5	10	10	21	21
5.21 to 5.40	5	10	10	20	20
5.41 to 5.60	5	10	10	19	19
5.61 to 5.70	5	9	9	19	19
5.71 to 5.80	5	9	9	18	18
5.81 to 5.90	4	9	9	18	18

Table 2-3. Category A					
	Total Number of Packages in Tare Sample				
	Note: Total number of packages to be opened for tare determination Numbers include those packages opened for initial tare sample				
Sample Size	12	24		48	
Initial Tare Sample Size	**2**	**2**	**3**	**2**	**3**
Ratio of R_c/R_t					
5.91 to 6.10	4	9	9	17	17
6.11 to 6.20	4	8	8	17	17
6.21 to 6.50	4	8	8	16	16
6.51 to 6.70	4	8	8	15	15
6.71 to 6.80	4	7	7	15	15
6.81 to 7.00	4	7	7	14	14
7.01 to 7.20	3	7	7	14	14
7.21 to 7.40	3	7	7	13	13
7.41 to 7.60	3	6	6	13	13
7.61 to 8.00	3	6	6	12	12
8.01 to 8.20	3	6	6	11	11
8.21 to 8.50	3	5	5	11	11
8.51 to 8.80	3	5	5	10	10
8.81 to 9.00	2	5	5	10	10
9.01 to 9.30	2	5	5	9	9
9.31 to 9.70	2	4	4	9	9
9.71 to 10.40	2	4	4	8	8
10.41 to 10.90	2	4	4	7	7
10.91 to 11.30	2	3	3	7	7
11.31 to 12.50	2	3	3	6	6
12.51 to 13.20	2	3	3	5	5
13.21 to 13.90	2	2	3	5	5
13.91 to 16.00	2	2	3	4	4
16.01 to 19.10	2	2	3	3	3
19.11 to 19.20	2	2	3	2	3
Initial Tare Sample Size	**2**	**2**	**3**	**2**	**3**

Table 2-4. Category B		
	Total Number of Packages in Tare Sample **Note:** Total number of packages to be opened for tare determination. Numbers include those packages opened for initial tare sample.	
Sample Size	**10**	**30**
Initial Tare Sample Size	**2**	**5**
Ratio of R_c/R_t		
If the ratio is zero, based on a "zero" range of tare, use Initial Tare Sample Size. If the ratio is "zero" based on a "zero" range of net weight, open all the packages in the sample.	2	5
If the ratio is greater than 0 but less than or equal to 0.2	10	30
0.21 to 0.40	10	29
0.41 to 0.60	10	28
0.61 to 0.80	9	26
0.81 to 1.00	8	24
1.01 to 1.20	8	23
1.21 to 1.40	7	21
1.41 to 1.60	7	19
1.61 to 1.80	6	17
1.81 to 2.00	5	15
2.01 to 2.20	5	14
2.21 to 2.40	5	13
2.41 to 2.60	4	12
2.61 to 2.80	4	11
2.81 to 3.00	4	10
3.01 to 3.20	3	9
3.21 to 3.60	3	8
3.61 to 3.80	3	7
3.81 to 4.40	2	6
If the ratio is greater than 4.40, use the Initial Tare Sample Size	2	5

Table 2-5. Maximum Allowable Variations (MAVs) for Packages Labeled by Weight

Do Not Use this Table for Meat and Poultry Products Subject to USDA Regulations – Use Table 2-9.
For Polyethylene Sheeting and Film, see Table 2-10. Exceptions to the MAVs.

Labeled Quantity	Maximum Allowable Variations
Less than 36 g, 0.08 lb, or 1.28 oz	10 % of labeled quantity
36 g or more to 54 g **0.08 lb or more to 0.12 lb** 1.28 oz or more to 1.92 oz	3.6 g **0.008 lb** $^1/_8$ oz
More than 54 g to 81 g **More than 0.12 lb to 0.18 lb** More than 1.92 oz to 2.88 oz	5.4 g **0.012 lb** $^3/_{16}$ oz
More than 81 g to 117 g **More than 0.18 lb to 0.26 lb** More than 2.88 oz to 4.16 oz	7.2 g **0.016 lb** $^1/_4$ oz
More than 117 g to 154 g **More than 0.26 lb to 0.34 lb** More than 4.16 oz to 5.44 oz	9.0 g **0.020 lb** $^5/_{16}$ oz
More than 154 g to 208 g **More than 0.34 lb to 0.46 lb** More than 5.44 oz to 7.36 oz	10.8 g **0.024 lb** $^3/_8$ oz
More than 208 g to 263 g **More than 0.46 lb to 0.58 lb** More than 7.36 oz to 9.28 oz	12.7 g **0.028 lb** $^7/_{16}$ oz
More than 263 g to 317 g **More than 0.58 lb to 0.70 lb** More than 9.28 oz to 11.20 oz	14.5 g **0.032 lb** $^1/_2$ oz
More than 317 g to 381 g **More than 0.70 lb to 0.84 lb** More than 11.20 oz to 13.44 oz	16.3 g **0.036 lb** $^9/_{16}$ oz
More than 381 g to 426 g **More than 0.84 lb to 0.94 lb** More than 13.44 oz to 15.04 oz	18.1 g **0.040 lb** $^5/_8$ oz
More than 426 g to 489 g **More than 0.94 lb to 1.08 lb** More than 15.04 oz to 17.28 oz	19.9 g **0.044 lb** $^{11}/_{16}$ oz
More than 489 g to 571 g More than 1.08 lb to 1.26 lb	21.7 g 0.048 lb
More than 571 g to 635 g More than 1.26 lb to 1.40 lb	23.5 g 0.052 lb
More than 635 g to 698 g More than 1.40 lb to 1.54 lb	25.4 g 0.056 lb

Table 2-5. Maximum Allowable Variations (MAVs) for Packages Labeled by Weight

Do Not Use this Table for Meat and Poultry Products Subject to USDA Regulations – Use Table 2-9.
For Polyethylene Sheeting and Film, see Table 2-10. Exceptions to the MAVs.

Labeled Quantity	Maximum Allowable Variations
More than 698 g to 771 g More than 1.54 lb to 1.70 lb	27.2 g 0.060 lb
More than 771 g to 852 g More than 1.70 lb to 1.88 lb	29.0 g 0.064 lb
More than 852 g to 970 g More than 1.88 lb to 2.14 lb	31.7 g 0.070 lb
More than 970 g to 1.12 kg More than 2.14 lb to 2.48 lb	35.3 g 0.078 lb
More than 1.12 kg to 1.25 kg More than 2.48 lb to 2.76 lb	39.0 g 0.086 lb
More than 1.25 kg to 1.45 kg More than 2.76 lb to 3.20 lb	42.6 g 0.094 lb
More than 1.45 kg to 1.76 kg More than 3.20 lb to 3.90 lb	49 g 0.11 lb
More than 1.76 kg to 2.13 kg More than 3.90 lb to 4.70 lb	54 g 0.12 lb
More than 2.13 kg to 2.63 kg More than 4.70 lb to 5.80 lb	63 g 0.14 lb
More than 2.63 kg to 3.08 kg More than 5.80 lb to 6.80 lb	68 g 0.15 lb
More than 3.08 kg to 3.58 kg More than 6.80 lb to 7.90 lb	77 g 0.17 lb
More than 3.58 kg to 4.26 kg More than 7.90 lb to 9.40 lb	86 g 0.19 lb
More than 4.26 kg to 5.30 kg More than 9.40 lb to 11.70 lb	99 g 0.22 lb
More than 5.30 kg to 6.48 kg More than 11.70 lb to 14.30 lb	113 g 0.25 lb
More than 6.48 kg to 8.02 kg More than 14.30 lb to 17.70 lb	127 g 0.28 lb
More than 8.02 kg to 10.52 kg More than 17.70 lb to 23.20 lb	140 g 0.31 lb
More than 10.52 kg to 14.33 kg More than 23.20 lb to 31.60 lb	167 g 0.37 lb
More than 14.33 kg to 19.23 kg More than 31.60 lb to 42.40 lb	199 g 0.44 lb

Table 2-5. Maximum Allowable Variations (MAVs) for Packages Labeled by Weight	
Do Not Use this Table for Meat and Poultry Products Subject to USDA Regulations – Use Table 2-9. For Polyethylene Sheeting and Film, see Table 2-10. Exceptions to the MAVs.	
Labeled Quantity	**Maximum Allowable Variations**
More than 19.23 kg to 24.67 kg More than 42.40 lb to 54.40 lb	226 g 0.50 lb
More than 24.67 kg More than 54.40 lb	2 % of labeled quantity

(Amended 2004)

Table 2-6. Maximum Allowable Variations for Packages Labeled by Liquid and Dry Volume	
Do Not Use this Table for Meat and Poultry Products Subject to USDA Regulations For Mulch, see Table 2-10. Exceptions to the Maximum Allowable Variations, Use Table 2-9 for USDA –Regulated Products.	
Labeled Quantity	**Maximum Allowable Variations (MAVs)**
3 mL or less **0.50 fl oz or less** 0.18 in^3 or less	0.5 mL **0.02 fl oz** 0.03 in^3
More than 3 mL to 8 mL More than 0.18 in^3 to 0.49 in^3	1.0 mL 0.06 in^3
More than 8 mL to 14 mL More than 0.49 in^3 to 0.92 in^3	1.5 mL 0.09 in^3
More than 14 mL to 22 mL **More than 0.50 fl oz to 0.75 fl oz** More than 0.92 in^3 to 1.35 in^3	1.7 mL **0.06 fl oz** 0.10 in^3
More than 22 mL to 66 mL **More than 0.75 fl oz to 2.25 fl oz** More than 1.35 in^3 to 4.06 in^3	3.8 mL **0.13 fl oz** 0.23 in^3
More than 66 mL to 125 mL **More than 2.25 fl oz to 4.25 fl oz** More than 4.06 in^3 to 7.66 in^3	5.6 mL **0.19 fl oz** 0.34 in^3
More than 125 mL to 170 mL **More than 4.25 fl oz to 5.75 fl oz** More than 7.66 in^3 to 10.37 in^3	7.3 mL **0.25 fl oz** 0.45 in^3
More than 170 mL to 221 mL **More than 5.75 fl oz to 7.50 fl oz** More than 10.37 in^3 to 13.53 in^3	9.1 mL **0.31 fl oz** 0.55 in^3
More than 221 mL to 347 mL **More than 7.50 fl oz to 11.75 fl oz** More than 13.53 in^3 to 21.20 in^3	11.2 mL **0.38 fl oz** 0.68 in^3

Table 2-6. Maximum Allowable Variations for Packages Labeled by Liquid and Dry Volume

Do Not Use this Table for Meat and Poultry Products Subject to USDA Regulations
For Mulch, see Table 2-10. Exceptions to the Maximum Allowable Variations.
Use Table 2-9 for USDA –Regulated Products.

Labeled Quantity	Maximum Allowable Variations (MAVs)
More than 347 mL to 502 mL **More than 11.75 fl oz to 17.00 fl oz** More than 21.20 in^3 to 30.67 in^3	14.7 mL **0.5 fl oz** 0.90 in^3
More than 502 mL to 621 mL More than 17 fl oz to 21 fl oz More than 30.67 in^3 to 37.89 in^3	18.6 mL **0.63 fl oz** 1.13 in^3
More than 621 mL to 798 mL More than 21 fl oz to 27 fl oz More than 37.89 in^3 to 48.72 in^3	22.1 mL **0.75 fl oz** 1.35 in^3
More than 798 mL to 916 mL More than 27 fl oz to 31 fl oz More than 48.72 in^3 to 55.94 in^3	26.0 mL **0.88 fl oz** 1.58 in^3
More than 916 mL to 1.15 L **More than 31 fl oz to 39 fl oz** More than 55.94 in^3 to 70.38 in^3	29 mL **1 fl oz** 1.80 in^3
More than 1.15 L to 1.62 L **More than 39 fl oz to 55 fl oz** More than 70.38 in^3 to 99.25 in^3	36 mL **1.25 fl oz** 2.25 in^3
More than 1.62 L to 2.04 L **More than 55 fl oz to 69 fl oz** More than 99.25 in^3 to 124.5 in^3	44 mL **1.5 fl oz** 2.70 in^3
More than 2.04 L to 2.51 L **More than 69 fl oz to 85 fl oz** More than 124.5 in^3 to 153.3 in^3	51 mL **1.75 fl oz** 3.1 in^3
More than 2.51 L to 3.04 L **More than 85 fl oz to 103 fl oz** More than 153.3 in^3 to 185.8 in^3	59 mL **2 fl oz** 3.6 in^3
More than 3.04 L to 4.73 L **More than 103 fl oz to 160 fl oz** More than 185.8 in^3 to 288.7 in^3	73 mL **2.5 fl oz** 4.5 in^3
More than 4.73 L to 5.48 L **More than 160 fl oz to 185.6 fl oz** More than 288.7 in^3 to 334.9 in^3	88 mL **3 fl oz** 5.4 in^3
More than 5.48 L to 7.09 L **More than 185.6 fl oz to 240 fl oz** More than 334.9 in^3 to 443.1 in^3	103 mL **3.5 fl oz** 6.3 in^3
More than 7.09 L to 8.04 L **More than 240 fl oz to 272 fl oz** More than 443.1 in^3 to 490.8 in^3	118 mL **4 fl oz** 7.2 in^3

Table 2-6. Maximum Allowable Variations for Packages Labeled by Liquid and Dry Volume	
Do Not Use this Table for Meat and Poultry Products Subject to USDA Regulations For Mulch, see Table 2-10. Exceptions to the Maximum Allowable Variations, Use Table 2-9 for USDA –Regulated Products.	
Labeled Quantity	**Maximum Allowable Variations (MAVs)**
More than 8.04 L to 10.17 L **More than 272 fl oz to 344 fl oz** More than 490.8 in^3 to 620.8 in^3	133 mL **4.5 fl oz** 8.1 in^3
More than 10.17 L to 11.59 L **More than 344 fl oz to 392 fl oz** More than 620.8 in^3 to 707.4 in^3	147 mL **5 fl oz** 9.0 in^3
More than 11.59 L to 16.56 L **More than 392 fl oz to 560 fl oz** More than 707.4 in^3 to 1 010 in^3	177 mL **6 fl oz** 10.8 in^3
More than 16.56 L to 18.92 L **More than 560 fl oz to 640 fl oz (5 gal)** More than 1 010 in^3 into 1 155 in^3	207 mL **7 fl oz** 12.6 in^3
More than 18.92 L to 23.65 L **More than 640 fl oz to 800 fl oz** More than 1 155 in^3 to 1 443 in^3	236 mL **8 fl oz** 14.4 in^3
More than 23.65 L to 26.73 L **More than 800 fl oz to 904 fl oz** More than 1 443 in^3 to 1 631 in^3	266 mL **9 fl oz** 16.2 in^3
More than 26.73 L **More than 904 fl oz** More than 1 631 in^3	**1 % of labeled quantity**

(Amended 2004)

| Table 2-7. Maximum Allowable Variations (MAVs) for Packages Labeled by Count ||
Labeled Quantity	Maximum Allowable Variations (MAVs)
17 or less	0
18 to 50	1
51 to 83	2
84 to 116	3
117 to 150	4
151 to 200	5
201 to 240	6
241 to 290	7
291 to 345	8
346 to 400	9
401 to 465	10
466 to 540	11
541 to 625	12
626 to 725	13
726 to 815	14
816 to 900	15
901 to 990	16
991 to 1075	17
1076 to 1165	18
1166 to 1250	19
1251 to 1333	20
1334 or more	1.5 % of labeled count rounded off to the nearest whole number

Table 2-8. Maximum Allowable Variations for Packages Labeled by Length, (Width), or Area For Textiles, Polyethylene Sheeting and Film – Use Table 2-10.	
Labeled Quantity	**Maximum Allowable Variations (MAVs) of Labeled Quantity**
1 m or less 1 yd or less	3 %
More than 1 m to 43 m More than 1 yd to 48 yd	1.5 %
More than 43 m to 87 m More than 48 yd to 96 yd	2 %
More than 87 m to 140 m More than 96 yd to 154 yd	2.5 %
More than 140 m to 301 m More than 154 yd to 330 yd	3 %
More than 301 m to 1 005 m More than 330 yd to 1 100 yd	4 %
More than 1 005 m or 1 100 yd	5 %
Maximum Allowable Variations for Packages Labeled by Area	
The MAV for packages labeled by area is 3 % of labeled quantity.	
For Textiles, Polyethylene Sheeting and Film, see Table 2-10. Exceptions to the MAVs.	

(Amended 2004)

Table 2-9. U.S. Department of Agriculture, Meat and Poultry Groups and Lower Limits for Individual Packages (Maximum Allowable Variations)		
Definition of Group and Labeled Quantity		**Lower Limit for Individual Weights (MAVs)**
Homogenous Fluid When Filled (e.g., baby food or containers of lard)	**All Other Products**	
Less than 85 g or 3 oz		10 % of labeled quantity
85 g or more to 453 g 3 oz or more to 16 oz		7.1 g 0.016 lb (0.25 oz)
More than 453 g More than 16 oz	85 g or more to 198 g 3 oz to 7 oz	14.2 g 0.031 lb (0.5 oz)
	More than 198 g to 1.36 kg 7 oz to 48 oz	28.3 g 0.062 lb (1 oz)
	More than 1.36 kg to 4.53 kg More than 48 oz to 160 oz	42.5 g 0.094 lb (1.5 oz)
	More than 4.53 kg More than 160 oz	1 % of labeled quantity

Table 2-10. Exceptions to the Maximum Allowable Variations for Textiles, Polyethylene Sheeting and Film, Mulch and Soil Labeled by Volume, Packaged Firewood, and Packages Labeled by Count with 50 Items or Fewer, and Specific Agricultural Seeds Labeled by Count.	
	Maximum Allowable Variations (MAVs)
Polyethylene Sheeting and Film	**Thickness** When the labeled thickness is 25 μm (1 mil or 0.001 in) or less, any individual thickness measurement of polyethylene film may be up to 35 % below the labeled thickness. When the labeled thickness is greater than 25 μm (1 mil or 0.001 in), individual thickness measurements of polyethylene sheeting may be up to 20 % less than the labeled thickness. The average thickness of a single package of polyethylene sheeting may be up to 4 % less than the labeled thickness. **Weight** The MAV for individual packages of polyethylene sheeting and film shall be 4 % of the labeled quantity.
Textiles	The MAVs are: For packages labeled with dimensions of 60 cm (24 in) or more: Three percent of the labeled quantity for negative errors and 6 % of the labeled quantity for plus errors. For packages labeled with dimensions less than 60 cm (24 in): 6 % of the labeled quantity for negative errors and 12 % for plus errors.
Mulch And Soil Labeled By Volume	The MAVs are: For individual packages: 5 % of the labeled volume. For example: One package may exceed the MAV for every 12 packages in the sample (e.g., when the sample size is 12 or fewer, 1 package may exceed the MAV and when the sample size is 48 packages, 4 packages may exceed the MAV).
Packaged Firewood and Packages Labeled by Count with 50 Items or Fewer	MAVs are not applied to these packages.

Table 2-10. Exceptions to the Maximum Allowable Variations for Textiles, Polyethylene Sheeting and Film, Mulch and Soil Labeled by Volume, Packaged Firewood, and Packages Labeled by Count with 50 Items or Fewer, and Specific Agricultural Seeds Labeled by Count.	
	Maximum Allowable Variations (MAVs)
Specific Agricultural Seeds Labeled By Count	The MAVS are: For corn seed: 2 % of the labeled count For soybean seed: 4 % of the labeled count For field bean seed: 5 % of the labeled count For wheat seed: 3 % of the labeled count

(Amended 2010)

Table 2-11. Accuracy Requirements for Packages Labeled by Low Count (50 or Fewer) and Packages Given Tolerances (Glass and Stemware)			
	1	2	3
Inspection Lot Size	Sample Size	For Packages Labeled by Low Count (50 or Fewer)	For Packages Given Tolerances (Glasses and Stemware)
		Number of Packages Allowed to Contain Less than the Labeled Count	Number of Package Errors that May Exceed the Allowable Difference
1 - 11	1-11	1	0
12 - 250	12	1	0
251 – 3200	24	2	1
More than 3200	48	3	2

(Amended 2004)

Appendix B. Random Number Tables

Reproduced from <u>*Million Random Digits*</u>, *used with permission of the Rand Corporation, Copyright, 1955, The Free Press* (http://www.rand.org/publications/classics/randomdigits)

All of the sampling plans presented in this handbook are based on the assumption that the packages constituting the sample are chosen at random from the inspection lot. Randomness in this instance means that every package in the lot has an equal chance of being selected as part of the sample. It does not matter what other packages have already been chosen, what the package net contents are, or where the package is located in the lot.

To obtain a random sample, two steps are necessary. First it is necessary to identify each package in the lot of packages with a specific number whether on the shelf, in the warehouse, or coming off the packaging line. Then it is necessary to obtain a series of random numbers. These random numbers indicate exactly which packages in the lot shall be taken for the sample.

The Random Number Table

The random number tables in Appendix B are composed of the digits from 0 through 9, with approximately equal frequency of occurrence. This appendix consists of 8 pages. On each page digits are printed in blocks of columns and blocks of rows. The printing of the table in blocks is intended only to make it easier to locate specific columns and rows.

Random Starting Place

Starting Page. The Random Digit pages are numbered B-2 through B-8. You can use the day of the week to determine the starting page or use the first page for the first lot you test in a location, the second page for the second lot, and so on, moving to the following page for each new lot.

Starting Column and Row. You may choose a starting page in the random number table and with eyes closed, drop a pencil anywhere on the page to indicate a starting place in the table.

For example, assume that testing takes place on the 3rd day of the week. Start with Table 3 of Appendix B. Assume you dropped your pencil on the page and it has indicated a starting place at column 22, row 45. That number is 1.

If one-digit random numbers are needed, record them, going down the column to the bottom of the page and then to the top of the next column, and so on. Ignore duplicates and record zero (0) as ten (10). Following on from the last example, these numbers are 3, 2, 9, 8, etc. If two-digit random numbers are needed, rule off the pages, and further pages if necessary, in columns of two digits each. If there is a single column left on the page, ignore this column, and rule the next page in columns of two. Again, ignore duplicate numbers and record 00 as 100. For example, using the same starting place as in the last example (Table 3, column 22, row 45), the recorded two-digit numbers would be 11, 34, 26, 95, etc. When three-digit numbers are needed, rule the page in columns of three. Record 000 as 1000. Starting on Table 3, column 22, row 45, the recorded numbers would be 119, 346, 269, 959, etc.

TABLE 1 – RANDOM NUMBER TABLES

11164	36318	75061	37674	26320	75100	10431	20418	19228	91792
21215	91791	76831	58678	87054	31687	93205	43685	19732	08468
10438	44482	66558	37649	08882	90870	12462	41810	01806	02977
36792	26236	33266	66583	60881	97395	20461	36742	02852	50564
73944	04773	12032	51414	82384	38370	00249	80709	72605	67497
49563	12872	14063	93104	78483	72717	68714	18048	25005	04151
64208	48237	41701	73117	33242	42314	83049	21933	92813	04763
51486	72875	38605	29341	80749	80151	33835	52602	79147	08868
99756	26360	64516	17971	48478	09610	04638	17141	09227	10606
71325	55217	13015	72907	00431	45117	33827	92873	02953	85474
65285	97198	12138	53010	94601	15838	16805	61004	43516	17020
17264	57327	38224	29301	31381	38109	34976	65692	98566	29550
95639	99754	31199	92558	68368	04985	51092	37780	40261	14479
61555	76404	86210	11808	12841	45147	97438	60022	12645	62000
78137	98768	04689	87130	79225	08153	84967	64539	79493	74917
62490	99215	84987	28759	19177	14733	24550	28067	68894	38490
24216	63444	21283	07044	92729	37284	13211	37485	10415	36457
16975	95428	33226	55903	31605	43817	22250	03918	46999	98501
59138	39542	71168	57609	91510	77904	74244	50940	31553	62562
29478	59652	50414	31966	87912	87154	12944	49862	96566	48825
96155	95009	27429	72918	08457	78134	48407	26061	58754	05326
29621	66583	62966	12468	20245	14015	04014	35713	03980	03024
12639	75291	71020	17265	41598	64074	64629	63293	53307	48766
14544	37134	54714	02401	63228	26831	19386	15457	17999	18306
83403	88827	09834	11333	68431	31706	26652	04711	34593	22561
67642	05204	30697	44806	96989	68403	85621	45556	35434	09532
64041	99011	14610	40273	09482	62864	01573	82274	81446	32477
17048	94523	97444	59904	16936	39384	97551	09620	63932	03091
93039	89416	52795	10631	09728	68202	20963	02477	55494	39563
82244	34392	96607	17220	51984	10753	76272	50985	97593	34320
96990	55244	70693	25255	40029	23289	48819	07159	60172	81697
09119	74803	97303	88701	51380	73143	98251	78635	27556	20712
57666	41204	47589	78364	38266	94393	70713	53388	79865	92069
46492	61594	26729	58272	81754	14648	77210	12923	53712	87771
08433	19172	08320	20839	13715	10597	17234	39355	74816	03363
10011	75004	86054	41190	10061	19660	03500	68412	57812	57929
92420	65431	16530	05547	10683	88102	30176	84750	10115	69220
35542	55865	07304	47010	43233	57022	52161	82976	47981	46588
86595	26247	18552	29491	33712	32285	64844	69395	41387	87195
72115	34985	58036	99137	47482	06204	24138	24272	16196	04393
07428	58863	96023	88936	51343	70958	96768	74317	27176	29600
35379	27922	28906	55013	26937	48174	04197	36074	65315	12537
10982	22807	10920	26299	23593	64629	57801	10437	43965	15344
90127	33341	77806	12446	15444	49244	47277	11346	15884	28131
63002	12990	23510	68774	48983	20481	59815	67248	17076	78910
40779	86382	48454	65269	91239	45989	45389	54847	77919	41105
43216	12608	18167	84631	94058	82458	15139	76856	86019	47928
96167	64375	74108	93643	09204	98855	59051	56492	11933	64958
70975	62693	35684	72607	23026	37004	32989	24843	01128	74658
85812	61875	23570	75754	29090	40264	80399	47254	40135	69916

TABLE 2 – RANDOM DIGITS

40603	16152	83235	37361	98783	24838	39793	80954	76865	32713
40941	53585	69958	60916	71018	90561	84505	53980	64735	85140
73505	83472	55953	17957	11446	22618	34771	25777	27064	13526
39412	16013	11442	89320	11307	49396	39805	12249	57656	88686
57994	76748	54627	48511	78646	33287	35524	54522	08795	56273
61834	59199	15469	82285	84164	91333	90954	87186	31598	25942
91402	77227	79516	21007	58602	81418	87838	18443	76162	51146
58299	83880	20125	10794	37780	61705	18276	99041	78135	99661
40684	99948	33880	76413	63839	71371	32392	51812	48248	96419
75978	64298	08074	62055	73864	01926	78374	15741	74452	49954
34556	39861	88267	76068	62445	64361	78685	24246	27027	48239
65990	57048	25067	77571	77974	37634	81564	98608	37224	49848
16381	15069	25416	87875	90374	86203	29677	82543	37554	89179
52458	88880	78352	67913	09245	47773	51272	06976	99571	33365
33007	85607	92008	44897	24964	50559	79549	85658	96865	24186
38712	31512	08588	61490	72294	42862	87334	05866	66269	43158
58722	03678	19186	69602	34625	75958	56869	17907	81867	11535
26188	69497	51351	47799	20477	71786	52560	66827	79419	70886
12893	54048	07255	86149	99090	70958	50775	31768	52903	27645
33186	81346	85095	37282	85536	72661	32180	40229	19209	74939
79893	29448	88392	54211	61708	83452	61227	81690	42265	20310
48449	15102	44126	19438	23382	14985	37538	30120	82443	11152
94205	04259	68983	50561	06902	10269	22216	70210	60736	58772
38648	09278	81313	77400	41126	52614	93613	27263	99381	49500
04292	46028	75666	26954	34979	68381	45154	09314	81009	05114
17026	49737	85875	12139	59391	81830	30185	83095	78752	40899
48070	76848	02531	97737	10151	18169	31709	74842	85522	74092
30159	95450	83778	46115	99178	97718	98440	15076	21199	20492
12148	92231	31361	60650	54695	30035	22765	91386	70399	79270
73838	77067	24863	97576	01139	54219	02959	45696	98103	78867
73547	43759	95632	39555	74391	07579	69491	02647	17050	49869
07277	93217	79421	21769	83572	48019	17327	99638	87035	89300
65128	48334	07493	28098	52087	55519	83718	60904	48721	17522
38716	61380	60212	05099	21210	22052	01780	36813	19528	07727
31921	76458	73720	08657	74922	61335	41690	41967	50691	30508
57238	27464	61487	52329	26150	79991	64398	91273	26824	94827
24219	41090	08531	61578	08236	41140	76335	91189	66312	44000
31309	49387	02330	02476	96074	33256	48554	95401	02642	29119
20750	97024	72619	66628	66509	31206	55293	24249	02266	39010
28537	84395	26654	37851	80590	53446	34385	86893	87713	26842
97929	41220	86431	94485	28778	44997	38802	56594	61363	04206
40568	33222	40486	91122	43294	94541	40988	02929	83190	74247
41483	92935	17061	78252	40498	43164	68646	33023	64333	64083
93040	66476	24990	41099	65135	37641	97613	87282	63693	55299
76869	39300	84978	07504	36835	72748	47644	48542	25076	68626
02982	57991	50765	91930	21375	35604	29963	13738	03155	59914
94479	76500	39170	06629	10031	48724	49822	44021	44335	26474
52291	75822	95966	90947	65031	75913	52654	63377	70664	60082
03684	03600	52831	55381	97013	19993	41295	29118	18710	64851
58939	28366	86765	67465	45421	74228	01095	50987	83833	37216

TABLE 3 – RANDOM DIGITS

37100	62492	63642	47638	13925	80113	88067	42575	44078	62703
53406	13855	38519	29500	62479	01036	87964	44498	07793	21599
55172	81556	18856	59043	64315	38270	25677	01965	21310	28115
40353	84807	47767	46890	16053	32415	60259	99788	55924	22077
18899	09612	77541	57675	70153	41179	97535	82889	27214	03482
68141	25340	92551	11326	60939	79355	41544	88926	09111	86431
51559	91159	81310	63251	91799	41215	87412	35317	74271	11603
92214	33386	73459	79359	65867	39269	57527	69551	17495	91456
15089	50557	33166	87094	52425	21211	41876	42525	36625	63964
96461	00604	11120	22254	16763	19206	67790	88362	01880	37911
28177	44111	15705	73835	69399	33602	13660	84342	97667	80847
66953	44737	81127	07493	07861	12666	85077	95972	96556	80108
19712	27263	84575	49820	19837	69985	34931	67935	71903	82560
68756	64757	19987	92222	11691	42502	00952	47981	97579	93408
75022	65332	98606	29451	57349	39219	08585	31502	96936	96356
11323	70069	90269	89266	46413	61615	66447	49751	15836	97343
55208	63470	18158	25283	19335	53893	87746	72531	16826	52605
11474	08786	05594	67045	13231	51186	71500	50498	59487	48677
81422	86842	60997	79669	43804	78690	58358	87639	24427	66799
21771	75963	23151	90274	08275	50677	99384	94022	84888	80139
42278	12160	32576	14278	34231	20724	27908	02657	19023	07190
17697	60114	63247	32096	32503	04923	17570	73243	76181	99343
05686	30243	34124	02936	71749	03031	72259	26351	77511	00850
52992	46650	89910	57395	39502	49738	87854	71066	84596	33115
94518	93984	81478	67750	89354	01080	25988	84359	31088	13655
00184	72186	78906	75480	71140	15199	69002	08374	22126	23555
87462	63165	79816	61630	50140	95319	79205	79202	67414	60805
88692	58716	12273	48176	86038	78474	76730	82931	51595	20747
20094	42962	41382	16768	13261	13510	04822	96354	72001	68642
60935	81504	50520	82153	27892	18029	79663	44146	72876	67843
51392	85936	43898	50596	81121	98122	69196	54271	12059	62539
54239	41918	79526	46274	24853	67165	12010	04923	20273	89405
57892	73394	07160	90262	48731	46648	70977	58262	78359	50436
02330	74736	53274	44468	53616	35794	54838	39114	68302	26855
76115	29247	55342	51299	79908	36613	68361	18864	13419	34950
63312	81886	29085	20101	38037	34742	78364	39356	40006	49800
27632	21570	34274	56426	00330	07117	86673	46455	66866	76374
06335	62111	44014	52567	79480	45886	92585	87828	17376	35254
64142	87676	21358	88773	10604	62834	63971	03989	21421	76086
28436	25468	75235	75370	63543	76266	27745	31714	04219	00699
09522	83855	85973	15888	29554	17995	37443	11461	42909	32634
93714	15414	93712	02742	34395	21929	38928	31205	01838	60000
15681	53599	58185	73840	88758	10618	98725	23146	13521	47905
77712	23914	08907	43768	10304	61405	53986	61116	76164	54958
78453	54844	61509	01245	91199	07482	02534	08189	62978	55516
24860	68284	19367	29073	93464	06714	45268	60678	58506	23700
37284	06844	78887	57276	42695	03682	83240	09744	63025	60997
35488	52473	37634	32569	39590	27379	23520	29714	03743	08444
51595	59909	35223	44991	29830	56614	59661	83397	38421	17503
90660	35171	30021	91120	78793	16827	89320	08260	09181	53616

TABLE 4 – RANDOM DIGITS

54723	56527	53076	38235	42780	22716	36400	48028	78196	92985
84828	81248	25548	34075	43459	44628	21866	90350	82264	20478
65799	01914	81363	05173	23674	41774	25154	73003	87031	94368
87917	38549	48213	71708	92035	92527	55484	32274	87918	22455
26907	88173	71189	28377	13785	87469	35647	19695	33401	51998
68052	65422	88460	06352	42379	55499	60469	76931	83430	24560
42587	68149	88147	99700	56124	53239	38726	63652	36644	50876
97176	55416	67642	05051	89931	19482	80720	48977	70004	03664
53295	87133	38264	94708	00703	35991	76404	82249	22942	49659
23011	94108	29196	65187	69974	01970	31667	54307	40032	30031
75768	49549	24543	63285	32803	18301	80851	89301	02398	99891
86668	70341	66460	75648	78678	27770	30245	44775	56120	44235
56727	72036	50347	33521	05068	47248	67832	30960	95465	32217
27936	78010	09617	04408	18954	61862	64547	52453	83213	47833
31994	69072	37354	93025	38934	90219	91148	62757	51703	84040
02985	95303	15182	50166	11755	56256	89546	31170	87221	63267
89965	10206	95830	95406	33845	87588	70237	84360	19629	72568
45587	29611	98579	42481	05359	36578	56047	68114	58583	16313
01071	08530	74305	77509	16270	20889	99753	88035	55643	18291
90209	68521	14293	39194	68803	32052	39413	26883	83119	69623
04982	68470	27875	15480	13206	44784	83601	03172	07817	01520
19740	24637	97377	32112	74283	69384	49768	64141	02024	85380
50197	79869	86497	68709	42073	28498	82750	43571	77075	07123
46954	67536	28968	81936	95999	04319	09932	66223	45491	69503
82549	62676	31123	49899	70512	95288	15517	85352	21987	08669
61798	81600	80018	84742	06103	60786	01408	75967	29948	21454
57666	29055	46518	01487	30136	14349	56159	47408	78311	25896
29805	64994	66872	62230	41385	58066	96600	99301	85976	84194
06711	34939	19599	76247	87879	97114	74314	39599	43544	36255
13934	46885	58315	88366	06138	37923	11192	90757	10831	01580
28549	98327	99943	25377	17628	65468	07875	16728	22602	33892
40871	61803	25767	55484	90997	86941	64027	01020	39518	34693
47704	38355	71708	80117	11361	88875	22315	38048	42891	87885
62611	19698	09304	29265	07636	08508	23773	56545	08015	28891
03047	83981	11916	09267	67316	87952	27045	62536	32180	60936
26460	50501	31731	18938	11025	18515	31747	96828	58258	97107
01764	25959	69293	89875	72710	49659	66632	25314	95260	22146
11762	54806	02651	52912	32770	64507	59090	01275	47624	16124
31736	31695	11523	64213	91190	10145	34231	36405	65860	48771
97155	48706	52239	21831	49043	18650	72246	43729	63368	53822
31181	49672	17237	04024	65324	32460	01566	67342	94986	36106
32115	82683	67182	89030	41370	50266	19505	57724	93358	49445
07068	75947	71743	69285	30395	81818	36125	52055	20289	16911
26622	74184	75166	96748	34729	61289	36908	73686	84641	45130
02805	52676	22519	47848	68210	23954	63085	87729	14176	45410
32301	58701	04193	30142	99779	21697	05059	26684	63516	75925
26339	56909	39331	42101	01031	01947	02257	47236	19913	90371
95274	09508	81012	42413	11278	19354	68661	04192	36878	84366
24275	39632	09777	98800	48027	96908	08177	15364	02317	89548
36116	42128	65401	94199	51058	10759	47244	99830	64255	40516

TABLE 5 – RANDOM DIGITS

47505	02008	20300	87188	42505	40294	04404	59286	95914	07191
13350	08414	64049	94377	91059	74531	56228	12307	87871	97064
33006	92690	69248	97443	38841	05051	33756	24736	43508	53566
55216	63886	06804	11861	30968	74515	40112	40432	18682	02845
21991	26228	14801	19192	45110	39937	81966	23258	99348	61219
71025	28212	10474	27522	16356	78456	46814	28975	01014	91458
65522	15242	84554	74560	26206	49520	65702	54193	25583	54745
27975	54923	90650	06170	99006	75651	77622	20491	53329	12452
07300	09704	36099	61577	34632	55176	87366	19968	33986	46445
54357	13689	19569	03814	47873	34086	28474	05131	46619	41499
00977	04481	42044	08649	83107	02423	46919	59586	58337	32280
13920	78761	12311	92808	71581	85251	11417	85252	61312	10266
08395	37043	37880	34172	80411	05181	58091	41269	22626	64799
46166	67206	01619	43769	91727	06149	17924	42628	57647	76936
87767	77607	03742	01613	83528	66251	75822	83058	97584	45401
29880	95288	21644	46587	11576	30568	56687	83239	76388	17857
36248	36666	14894	59273	04518	11307	67655	08566	51759	41795
12386	29656	30474	25964	10006	86382	46680	93060	52337	56034
52068	73801	52188	19491	76221	45685	95189	78577	36250	36082
41727	52171	56719	06054	34898	93990	89263	79180	39917	16122
49319	74580	57470	14600	22224	49028	93024	21414	90150	15686
88786	76963	12127	25014	91593	98208	27991	12539	14357	69512
84866	95202	43983	72655	89684	79005	85932	41627	87381	38832
11849	26482	20461	99450	21636	13337	55407	01897	75422	05205
54966	17594	57393	73267	87106	26849	68667	45791	87226	74412
10959	33349	80719	96751	25752	17133	32786	34368	77600	41809
22784	07783	35903	00091	73954	48706	83423	96286	90373	23372
86037	61791	33815	63968	70437	33124	50025	44367	98637	40870
80037	65089	85919	74391	36170	82988	52311	59180	37846	98028
72751	84359	15769	13615	70866	37007	74565	92781	37770	76451
18532	03874	66220	79050	66814	76341	42452	65365	07167	90134
22936	22058	49171	11027	07066	14606	11759	19942	21909	15031
66397	76510	81150	00704	94990	68204	07242	82922	65745	51503
89730	23272	65420	35091	16227	87024	56662	59110	11158	67508
81821	75323	96068	91724	94679	88062	13729	94152	59343	07352
94377	82554	53586	11432	08788	74053	98312	61732	91248	23673
68485	49991	53165	19865	30288	00467	98105	91483	89389	61991
07330	07184	86788	64577	47692	45031	36325	47029	27914	24905
10993	14930	35072	36429	26176	66205	07758	07982	33721	81319
20801	15178	64453	83357	21589	23153	60375	63305	37995	66275
79241	35347	66851	79247	57462	23893	16542	55775	06813	63512
43593	39555	97345	58494	52892	55080	19056	96192	61508	23165
29522	62713	33701	17186	15721	95018	76571	58615	35836	66260
88836	47290	67274	78362	84457	39181	17295	39626	82373	10883
65905	66253	91482	30689	81313	01343	37188	37756	04182	19376
44798	69371	07865	91756	42318	63601	53872	93610	44142	89830
35510	99139	32031	27925	03560	33806	85092	70436	94777	57963
50125	93223	64209	49714	73379	89975	38567	44316	60262	10777
25173	90038	63871	40418	23818	63250	05118	52700	92327	55449
68459	90094	44995	93718	83654	79311	18107	12557	09179	28416

TABLE 6 – RANDOM DIGITS

96195	07059	13266	31389	87612	88004	31843	83469	22793	14312
22408	94958	19095	58035	43831	32354	83946	57964	70404	32017
53896	23508	16227	56929	74329	12264	26047	66844	47383	42202
22565	02475	00258	79018	70090	37914	27755	00872	71553	56684
49438	20772	60846	69732	07612	70474	46483	21053	95475	53448
65620	34684	00210	04863	01373	19978	61682	69315	46766	83768
20246	26941	41298	04763	19769	25865	95937	03545	93561	73871
09433	09167	35166	32731	73299	41137	37328	28301	61629	05040
95552	73456	16578	88140	80059	50296	07656	01396	83099	09718
76053	05150	69125	69442	16509	03495	26427	58780	27576	31342
34822	35843	78468	82380	52313	71070	71273	10768	86101	51474
07753	04073	58520	80022	28185	16432	86909	82347	10548	83929
04204	94434	62798	81902	29977	57258	87826	35003	46449	76636
96770	19440	29700	42093	64369	69176	29732	37389	34054	28680
65989	62843	10917	34458	81936	84775	39415	10622	36102	16753
06644	94784	66995	61812	54215	01336	75887	57685	66114	76984
88950	46077	34651	12038	87914	20785	39705	73898	12318	78334
21482	95422	02002	33671	46764	50527	46276	77570	68457	62199
55137	61039	02006	69913	11291	87215	89991	26003	55271	08153
98441	81529	59607	65225	49051	28328	85535	37003	87211	10204
57168	30458	23892	07825	53447	53511	09315	42552	43135	57892
71886	65334	38013	09379	83976	42441	14086	33197	82671	05037
40418	59504	52383	07232	14179	59693	37668	26689	93865	78925
28833	76661	47277	92935	63193	94862	60560	72484	29755	40894
37883	62124	62199	49542	55083	20575	44636	92282	52105	77664
44882	33592	66234	13821	86342	00135	87938	57995	34157	99858
19082	13873	07184	21566	95320	28968	31911	06288	77271	76171
45316	29283	89318	55806	89338	79231	91545	55477	19552	03471
22788	55433	31188	74882	44858	69655	08096	70982	61300	23792
08293	86193	05026	21255	63082	92946	28748	25423	45282	57821
29223	70541	67115	84584	10100	33854	26466	77796	70698	99393
22681	80110	31595	09246	39147	11158	43298	36220	88841	11271
74580	90354	43744	22178	38084	60027	24201	71686	59767	33274
69093	71364	08107	96952	50005	30297	97417	89575	04676	35616
40456	91234	58090	65342	95002	28447	21'700	43137	13746	85959
72927	67349	83962	58912	59734	76323	02913	46306	53956	38936
61869	33093	81129	06481	89281	83629	81960	63704	56329	10357
40048	16520	07638	10797	22270	57350	72214	36410	95526	87614
68773	97669	28656	89938	12917	25630	08068	19445	76250	24727
09774	30751	49740	11385	91468	28900	76804	52460	52320	70493
46139	36689	82587	13586	35061	76128	38568	62300	43439	53434
26566	95323	32993	89988	12152	01862	93113	33875	31730	62941
06765	57141	48617	18282	13086	76064	83334	70192	15972	80429
35384	90380	12317	89702	33091	68835	62960	38010	52710	87604
49333	78482	36199	11355	86044	88760	03724	22927	91716	92332
45595	14044	56806	99126	85584	87750	78149	22723	48245	78126
79819	15054	76174	12206	06886	06814	43285	20008	75345	19779
11971	62234	74857	46401	20817	57591	41189	49604	29604	30660
11452	89318	53084	21993	62471	74101	61217	76536	58393	63718
38746	81271	96260	98137	60275	22647	33103	50090	29395	10016

TABLE 7 – RANDOM DIGITS

93369	13044	69686	78162	29132	51544	17925	56738	32683	83153
19360	55049	94951	76341	38159	31008	41476	05278	03909	02299
47798	89890	06893	65483	97658	74884	38611	27264	26956	83504
69223	32007	03513	61149	66270	73087	16795	76845	44645	44552
34511	50721	84850	34159	38985	75384	22965	55366	81632	78872
54031	59329	58963	52220	76806	98715	67452	78741	58128	00077
66722	85515	04723	92411	03834	12109	85185	37350	93614	15351
71059	07496	38404	18126	37894	44991	45777	02070	38159	23930
45478	86066	31135	33243	01190	47277	55146	56130	70117	83203
97246	91121	89437	20393	76598	99458	76665	83793	37448	32664
22982	25936	96417	34845	28942	65569	38253	77182	12996	19505
48243	62993	47132	85248	79160	90981	71696	79609	33809	60839
93514	14915	67960	82203	22598	94802	75332	95585	69542	79924
69707	98303	93069	16216	01542	51771	16833	20922	94415	27617
87467	91794	70814	12743	17543	04057	71231	11309	32780	83270
81006	81498	59375	30502	44868	81279	23585	49678	70014	10523
15458	83481	50187	43375	56644	72076	59403	65469	74760	69509
33469	12510	23095	48016	22064	39774	07373	10555	33345	21787
67198	07176	65996	18317	83083	11921	06254	68437	59481	54778
58037	92261	85504	55690	63488	26451	43223	38009	50567	09191
84983	68312	25519	56158	22390	12823	92390	28947	36708	25393
35554	02935	72889	68772	79774	14336	50716	63003	86391	94074
04368	17632	50962	71908	13105	76285	31819	16884	11665	16594
81311	60479	69985	30952	93067	70056	55229	83226	22555	66447
03823	89887	55828	74452	21692	55847	15960	47521	27784	25728
80422	65437	38797	56261	88300	35980	56656	45662	29219	49257
61307	49468	43344	43700	14074	19739	03275	99444	62545	23720
83873	82557	10002	80093	74645	33109	15281	38759	09342	69408
38110	16855	28922	93758	22885	36706	92542	60270	99599	17983
43892	91189	87226	56935	99836	85489	89693	49475	31941	78065
93683	09664	53927	49885	94979	88848	42642	93218	80305	49428
32748	02121	11972	96914	83264	89016	45140	20362	63242	86255
49211	92963	38625	65312	52156	36400	67050	64058	45489	24165
63365	64224	69475	57512	85097	05054	88673	96593	00902	53320
63576	26373	44610	43748	90399	06770	71609	90916	69002	57180
41078	47036	65524	68466	77613	20076	71969	47706	22506	81053
70846	89558	64173	15381	67322	70097	82363	90767	17879	32697
68800	64492	20162	32707	69510	82465	26821	79917	34615	35820
44977	89525	51269	63747	30997	97213	53016	65909	05723	50168
79354	63847	24395	53679	07667	67993	24634	78867	78516	00448
14954	22299	40156	52685	19093	06090	23800	06739	76836	19050
01711	98439	09446	33937	98956	85676	89493	05132	45886	49379
62328	55328	45738	93940	15772	81975	91017	21387	57949	13992
73004	62109	81907	71077	50322	66093	79921	61412	18347	21115
34218	89445	03609	52336	19005	15179	94958	99448	11612	76981
99159	01968	45886	86875	05196	64297	59339	39878	61548	56442
92858	29949	15817	93372	34732	61584	72007	58597	43802	51066
27396	97477	65554	71601	01540	26509	19487	39684	18676	41219
37103	45309	30129	43380	66638	10841	77292	40288	25826	61431
57347	97012	48428	20606	54138	75716	23741	50462	13221	47216

Appendix C. Model Inspection Report Forms

THIS PAGE INTENTIONALLY LEFT BLANK

Date:	**Random Package Report**			Sampling Plan: □ A □ B		Report Number:

Location (name, address):	Product/Brand Identity:	Manufacturer:	Container Description:
	Lot Codes:		

1. Labeled Quantity: (Enter weight for each package in Column 1 below.)	2. Unit of Measure:	3. MAV: (Look up the MAV for each package with a minus error (−), convert it to dimensionless units and enter this value in Column 4 below.)	5. Inspection Lot Size:	6. Sample Size (n):

7. Initial Tare Sample Size:	8. Number of MAVs Allowed:	9. Range of Package Errors (R_c):	10. Range of Tare Weights (R_t):	11. R_c/R_t (Box 9 ÷ Box 10 =):	12. Total No. of Tare Samples:

13. Avg. Tare Wt: □ Used Dry Tare □ Wet Tare □ Unused Dry Tare	13a. □ Tare Correction □ Moisture Allowance □ Not Applicable	14. Nominal Gross Wt: (Labeled Wt + Box 13 − Box 13a =)

	Pkg 1	Pkg 2	Pkg 3	Pkg 4	Pkg 5	Pkg 6	Pkg 7	Pkg 8	Pkg 9	Pkg 10
a. Gross Wt										
b. Tare Wt										
c. Net Wt										
d. Package Error (= Box a − Box 14)										

Product Description, Lot Code, Unit Price	Money Errors		Column 1 Labeled Net Weight	Package Errors		Column 4 MAV
	−	+		−	+	
1.						
2.						
3.						
4.						
5.						
6.						
7.						
8.						
9.						
10.						
11.						
12.						
13.						
14.						
15.						
16.						
17.						
			Totals			

15. Total Error:	16. Number of unreasonable minus (−) errors: (Compare each package error with the MAV in Column 4.)	17. Is Box 16 greater than Box 8? □ Yes, lot fails □ No, go to Box 18	18. Avg. error in dimensionless units: (Box 15 ÷ Box 6 =)	19. Avg. error in labeled units (Box 18 x Box 2 =)

20. Does Box 18 = zero (0) or Plus (+)? □ Yes, lot passes, go to Box 25 □ No, go to Box 21	21. Compute Sample Standard Deviation	22. Sample Correction Factor	23. Compute Sample Error Limit (Box 21 x Box 22 =)

24. Disregarding the signs, is Box 18 larger than Box 23? □ Yes, lot fails, go to Box 25 □ No, lot passes, go to Box 25	25. Disposition of Inspection Lot □ Approved □ Rejected

Comments	Official's Signature:
	Acknowledgement of Report:

Date: *January 20, 2010*	**Random Package Report – Example**		Sampling Plan: ☐ A ☐ B	Report Number: 17

Location (name, address): *L&O Market* *MacCorkle Ave* *Charleston, WV 25171*	Product/Brand Identity: *Ground Chuck* Lot Codes: *1, 19, 99*	Manufacturer: *Meat Dept - L&O Market*	Container Description: *2S Tray w/soaker and* *plastic wrap*

1. Labeled Quantity: (Enter weight for each package in Column 1 below.)	2. Unit of Measure: *0.001 lb*	3. MAV: (Look up the MAV for each package with a minus error (−), convert it to dimensionless units and enter this value in Column 4 below.)	5. Inspection Lot Size: 23	6. Sample Size (n): *12*
7. Initial Tare Sample Size: *2*	8. Number of MAVs Allowed: *0*	9. Range of Package Errors (R_c): *10*	10. Range of Tare Weights (R_t): *1*	11. R_c/R_t (Box 9 ÷ Box 10 =): *10*	12. Total No. of Tare Samples: *2*

13. Avg. Tare Wt: *0.0205 lb* ☐ Used Dry Tare ☐ Wet Tare ☐ Unused Dry Tare	13a. ☐ Tare Correction ☐ Moisture Allowance ☐ Not Applicable	14. Nominal Gross Wt: (Labeled Wt + Box 13 − Box 13a =) *Label Wt + 0.020 lb*

	Pkg 1	Pkg 2	Pkg 3	Pkg 4	Pkg 5	Pkg 6	Pkg 7	Pkg 8	Pkg 9	Pkg 10
a. Gross Wt	*1.852 lb*	*1.223 lb*								
b. Tare Wt	*0.020 lb*	*0.021 lb*								
c. Net Wt	*1.832 lb*	*1.202 lb*								
d. Package Error (= Box a − Box 14)	*−18*	*−8*								

Product Description, Lot Code, Unit Price	Money Errors		Column 1 Labeled Net Weight	Package Errors		Column 4 MAV
	−	+		−	+	
1. *Ground Chuck - 1, 19, 99 - $1.79 per lb*			*1.85 lb*	*18*		
2.			*1.21 lb*	*7*		
3.			*1.56 lb*	*8*		
4.			*1.98 lb*	*14*		
5.	*$0.04*		*1.07 lb*	*23*		*44*
6.			*1.55 lb*	*16*		
7.			*1.02 lb*	*2*		
8.	*$0.04*		*1.44 lb*	*25*		*56*
9.			*1.33 lb*	*16*		
10.			*2.03 lb*	*20*		*70*
11.			*1.73 lb*	*14*		
12.			*1.16 lb*	*11*		
13.						
14.						
15.						
16.						
17.			Totals			

15. Total Error: *−174*	16. Number of unreasonable minus (−) errors: (Compare each package error with the MAV in Column 4.) *0*	17. Is Box 16 greater than Box 8? ☐ Yes, lot <u>fails</u> ☐ No, go to Box 18	18. Avg. error in dimensionless units: (Box 15 ÷ Box 6 =) *−14.5*	19. Avg. error in labeled units (Box 18 x Box 2 =) *− 0.014 lb*
20. Does Box 18 = Zero (0) or Plus (+)? ☐ Yes, lot passes, go to Box 25 ☐ No, go to Box 21	21. Compute Sample Standard Deviation *6.721*	22. Sample Correction Factor *0.635*	23. Compute Sample Error Limit (Box 21 x Box 22 =) *4.267*	
24. Disregarding the signs, is Box 18 larger than Box 23? ☐ Yes, lot <u>fails</u>, go to Box 25 ☐ No, lot <u>passes</u>, go to Box 25	25. Disposition of Inspection Lot ☐ Approved ☐ Rejected			
Comments	Official's Signature: Acknowledgement of Report:			

Date:	**Standard Package Report**	Sampling Plan: ☐ A ☐ B	Report Number:

Location (name, address)	Product/Brand Identity	Manufacturer	Container Description
	Lot Codes		

1. Labeled Quantity:	2. Unit of Measure:	3. MAV:	4. MAV (dimensionless units) (Box 3 ÷ Box 2 =)	5. Inspection Lot Size:	6. Sample Size (n):

7. Initial Tare Sample Size:	8. Number of MAVs Allowed:	9. Range of Package Errors (R_e):	10. Range of Tare Weights (R_t):	11. R_e/R_t: (Box 9 ÷ 10 =)	12. Total Number of Tare Samples:

13. Average Tare Wt: ☐ Used Dry Tare ☐ Wet Tare ☐ Unused Dry Tare	13a. ☐ Tare Correction ☐ Moisture Allowance ☐ Vacuum Pack ☐ Not Applicable	14. Nominal Gross Wt: (Box 1 + Box13 − Box 13a =)

	Pkg 1	Pkg 2	Pkg 3	Pkg 4	Pkg 5	Pkg 6	Pkg 7	Pkg 8	Pkg 9	Pkg 10
a. Gross Wt										
b. Tare Wt										
c. Net Wt										

−	+	−	+	−	+	−	+
1.		13.		25.		37.	
2.		14.		26.		38.	
3.		15.		27.		39.	
4.		16.		28.		40.	
5.		17.		29.		41.	
6.		18.		30.		42.	
7.		19.		31.		43.	
8.		20.		32.		44.	
9.		21.		33.		45.	
10.		22.		34.		46.	
11.		23.		35.		47.	
12.		24.		36.		48.	
Total:	Total:	Total:	Total:	Total:	Total:	Total:	Total:

15. Total Error:	16. Number of unreasonable minus (−) errors (compare each package error with Box 4)	17. Is Box 16 greater than Box 8? ☐ Yes, lot fails ☐ No, go to Box 18	18. Average error in dimensionless units (Box 15 ÷ Box 6 =)	19. Average error in labeled units: (Box 18 x Box 2 =)

20. Does Box 18 = Zero (0) or Plus (+)? ☐ Yes, lot passes, go to Box 25 ☐ No, go to Box 21	21. Compute Sample Standard Deviation	22. Sample Correction Factor	23. Compute Sample Error Limit (Box 21 x Box 22 =)

24. Disregarding the signs, is Box 18 larger than Box 23? ☐ Yes, lot fails, go to Box 25 ☐ No, lot passes, go to Box 25	25. Disposition of Inspection Lot ☐ Approved ☐ Rejected

Comments:	Official's Signature
	Acknowledgement of Report

131

Date: *January 20, 2010*	**Standard Package Report – Example**		Sampling Plan: ☐ A ☐ B		Report Number: *16*

Location (name, address) *Volunteer Market* *18765 Alcoa Highway* *Knoxville, TN 37920*	Product/Brand Identity *Community Group Cookies (Thin Mints)* Lot Codes *April 2009 A & B*	Manufacturer *ABC Cookies Inc* *1069 Capitol Avenue* *Nashville, TN 37204*	Container Description *Cardboard Box/* *Plastic Liner*

1. Labeled Quantity: *453 g (1 lb)*	2. Unit of Measure: *0 001 lb*	3. MAV: *0 044 lb*	4. MAV (dimensionless units) (Box 3 ÷ Box 2 =) *44*	5. Inspection Lot Size: *172*	6. Sample Size (n): *12*

7. Initial Tare Sample Size: *2*	8. Number of MAVs Allowed: *0*	9. Range of Package Errors (R$_c$): *24*	10. Range of Tare Weights (R$_t$): *2*	11. R$_c$/R$_t$: (Box 9 ÷ 10 =) *12*	12. Total Number. of Tare Samples: *2*

13. Average Tare Wt: *0 014 lb* ☑ Used Dry Tare ☐ Wet Tare ☐ Unused Dry Tare	13a. ☐ Tare Correction ☐ Moisture Allowance ☐ Vacuum Pack ☐ Not Applicable	14. Nominal Gross Wt: (Box 1 + Box13 − Box 13a =) *1 014 lb*

	Pkg 1	Pkg 2	Pkg 3	Pkg 4	Pkg 5	Pkg 6	Pkg 7	Pkg 8	Pkg 9	Pkg 10
a. Gross Wt	*1 052 lb*	*1 026 lb*								
b. Tare Wt	*0 015 lb*	*0 013 lb*								
c. Net Wt	*1 037 lb*	*1 013 lb*								

−	+	−	+	−	+	−	+
1.	*38*			25.		37.	
2.	*12*	14.		26.		38.	
3.	*8*	15.		27.		39.	
4.	*4*	16.		28.		40.	
5. *3*		17.		29.		41.	
6. *2*		18.		30.		42.	
7.	*12*	19.		31.		43.	
8. *2*		20.		32.		44.	
9.	*4*	21.		33.		45.	
10. *1*		22.		34.		46.	
11. *0*		23.		35.		47.	
12.	*6*	24.		36.		48.	
Total: *9*	Total: *84*	Total:	Total:	Total:	Total:	Total:	Total:

15. Total Error: *+ 75*	16. Number of unreasonable minus (−) errors (compare each package error with Box 4) *0*	17. Is Box 16 greater than Box 8? ☐ Yes, lot <u>fails</u> ☑ No, go to Box 18	18. Average error in dimensionless units (Box 15 ÷ Box 6 =) *+ 6 25*	19. Average error in labeled units: (Box 18 x Box 2 =) *+ 0 006 lb*

20. Does Box 18 = Zero (0) or Plus (+)? ☑ Yes, lot <u>passes</u>, go to Box 25 ☐ No, go to Box 21	21. Compute Sample Standard Deviation	22. Sample Correction Factor	23. Compute Sample Error Limit (Box 21 x Box 22 =)

24. Disregarding the signs, is Box 18 larger than Box 23? ☐ Yes, lot <u>fails</u>, go to Box 25 ☐ No, lot <u>passes</u>. go to Box 25	25. Disposition of Inspection Lot ☑ Approved ☐ Rejected

Comments: *Lot Passes*	Official's Signature Acknowledgement of Report

Ice Glazed Package Worksheet

STEP

1. Package Price (if standard pack) $ _____ Price Per Pound (if random pack) $ _____

   Lot Size: _____ Sample Size: _____ Unit of Measure: _____

2. Number each package. Weigh each package for Gross Package Weight and enter Row 1.

3. Enter Labeled Net Weight in Row 2. (If dual units determine the larger unit.) _____

4. Record the Maximum Allowable Variation (MAV) in Row 3.

5. Weigh the receiving pan = _____ (enter in Row 4). (Clean and dry the receiving pan and verify the weight after each use. Thoroughly clean the sieve.)

6. Deglaze the product. Remove each package from the low temperature storage. Open the package immediately and place the contents in the sieve or other draining device (e.g., colander) under a gentle spray of cold water. Carefully agitate the product. Handle with care to avoid breaking the product. Continue the spraying process until all ice glaze that is seen or felt is removed.

7. Without shifting the product, incline the sieve to an angle of 17° to 20° (incline to facilitate drainage) and drain for 2 minutes using a stopwatch.

8. Immediate transfer the entire product to the receiving pan to determine the net weight.

9. To calculate the net weight (receiving pan and product) – (receiving pan) = Net Weight (enter in Row 5)

10. Calculate ± Package error (net weight [Row 5] – labeled net weight [Row 2]) = ± Error, (enter in Row 6).

Row	Package	1	2	3	4	5	6	7	8	9	10	11	12
1	Gross Pkg. Weight												
2	Labeled Net Weight												
3	MAV												
4	Receiving Pan Weight (step 5)												
5	Net Weight (step 9)												
6	± Error (step 10)												

Used Dry Tare _____

Transfer data from the "Ice Glazed Package Worksheet" to the "Ice Glazed Package Report"

(Added 2010)

133

Ice Glazed Package Worksheet – Example

<u>STEP</u>

1. Package Price (if standard pack) $ _6 99_   Price Per Pound (if random pack) $ _____

    Lot Size: ___6___ Sample Size: ___6___ Unit of Measure: ___0 001 lb___

2. Number each package. Weigh each package for Gross Package Weight and enter Row 1.

3. Enter Labeled Net Weight in Row 2. (If dual units determine the larger unit.) _1 lb/453 g_

4. Record the Maximum Allowable Variation (MAV) in Row 3.

5. Weigh the receiving pan = _0 795 lb_ (enter in Row 4). (Clean and dry the receiving pan and verify the weight after each use. Thoroughly clean the sieve.)

6. Deglaze the product. Remove each package from the low temperature storage. Open the package immediately and place the contents in the sieve or other draining device (e.g., colander) under a gentle spray of cold water. Carefully agitate the product. Handle the product with care to avoid breaking the product. Continue the spraying process until all ice glaze that is seen or felt is removed.

7. Without shifting the product, incline the sieve to an angle of 17° to 20° (incline to facilitate drainage) and drain for 2 minutes using a stopwatch.

8. Immediately transfer the entire product to the receiving pan to determine the net weight.

9. To calculate the net weight (receiving pan and product) – (receiving pan) = Net Weight (enter in Row 5)

10. Calculate ± Package error (net weight [Row 5] – labeled net weight [Row 2]) = ± Error, (enter in Row 6).

Row	Package	1	2	3	4	5	6	7	8	9	10	11	12
1	Gross Pkg. Weight	1.180	1.205	1.110	1.150	1.000	1.210						
2	Labeled Net Weight	1.000	1.000	1.000	1.000	1.000	1.000						
3	MAV	0.044	0.044	0.044	0.044	0.044	0.044						
4	Receiving Pan Weight (step 5)	0.795	0.795	0.795	0.795	0.795	0.795						
5	Net Weight (step 9)	0.985	0.975	1.000	1.030	0.930	0.980						
6	± Error (step 10)	- 0.015	- 0.025	0	+ 0.030	- 0.070	- 0.020						

Used Dry Tare _0 025 lb_

Transfer data from the "Ice Glazed Package Worksheet" to the "Ice Glazed Package Report"
(Added 2010)

Date:	**Ice Glazed Package Report**		Sampling Plan: ☐ A ☐ B	Report Number:

Location (name, address):	Product/Brand Identity:	Manufacturer:	Container Description:
	Lot Codes:		

1. Standard Pack Labeled Quantity: (If random packed, enter weight for each package in Column 1 below.)	2. Unit of Measure:	3. MAV: Look up the MAV for each package with a minus (−) error, enter value in Column 4.	5. Inspection Lot Size	6. Sample Size (n)

7. Price per lb: 7a. Standard Pack: Package Price _____ divide by (Box 1) = _____ 7b. Random Pack: Labeled Price per lb _____	8. No. of MAVs Allowed

	Pkg 1	Pkg 2	Pkg 3	Pkg 4	Pkg 5	Pkg 6	Pkg 7	Pkg 8	Pkg 9	Pkg 10	Pkg 11	Pkg 12
Pkg. Gross Wt												
a. Labeled Net Wt												
b. Gross: Rec. Pan & deglazed product Wt												
c. Tare: Rec. Pan Wt												
d. Net Wt (Box b − Box c=)												
e. Package Error (Box d − Box a =)												

Package #	Column 1 Labeled Net Weight (random pack only)	Package Errors −	Package Errors +	Column 4. MAV
1				
2				
3				
4				
5				
6				
7				
8				
9				
10				
11				
12				
Totals		f.	g.	

9. Total Error (add row e or Box f + g):	10. Number of unreasonable minus (−) errors (compare each package error with the MAV in Column 4):	11. Is Box 10 greater than Box 8? ☐ Yes, lot fails ☐ No, go to Box 12	12. Avg. error (Box 9 ÷ Box 6 =):

13. Does Box 12 = Zero (0) or Plus (+)? ☐ Yes, lot passes, go to Box 18 ☐ No, go to Box 14	14. Compute Sample Standard Deviation:	15. Sample Correction Factor:	16. Compute Sample Error Limit (Box 14 x Box 15 =)

17. Disregarding the signs, is Box 12 larger than Box 16? ☐ Yes, lot fails, go to Box 18 ☐ No, lot passes, go to Box 18	18. Disposition of Inspection Lot ☐ Approved ☐ Rejected	19. Economic Impact: (Box 12 x Box 7 x Box 5 =)

Comments:	Official's Signature:
	Acknowledgement of Report:

Date: *January 20, 2010*	**Ice Glazed Package Report – Example**			Sampling Plan: ☐ A ☐ B		Report Number: *103*

Location (name, address): *Ocean Fresh Market* *101 8th Street* *Key West, FL*	Product/Brand Identity: *Raw/Peeled Shrimp 71 – 90 Count* Lot Codes:	Manufacturer: *Ocean Fresh*	Container Description: *Plastic*

1. Standard Pack Labeled Quantity: *453 g (1 lb)* (If random packed, enter weight for each package in Column 1 below.)	2. Unit of Measure: *0 001 lb*	3. MAV: Look up the MAV for each package with a minus (−) error, enter value in Column 4. *0 044 lb*	5. Inspection Lot Size *6*	6. Sample Size (n) *6*

7. Price per lb: 7a. Standard Pack: Package Price $ *6 99* divide by (Box 1) = $ *6 99* 7b. Random Pack: Labeled Price per lb _____	8. No. of MAVs Allowed *0*

	Pkg 1	Pkg 2	Pkg 3	Pkg 4	Pkg 5	Pkg 6	Pkg 7	Pkg 8	Pkg 9	Pkg 10	Pkg 11	Pkg 12
Pkg. Gross Wt	*1.180*	*1.205*	*1.100*	*1.150*	*1.000*	*1.210*						
a. Labeled Net Wt	*1.000*	*1.000*	*1.000*	*1.000*	*1.000*	*1.000*						
b. Gross: Rec. Pan & deglazed product Wt												
c. Tare: Rec. Pan Wt	*0.795*	*0.795*	*0.795*	*0.795*	*0.795*	*0.795*						
d. Net Wt (Box b − Box c=)	*0.985*	*0.975*	*1.000*	*1.030*	*0.930*	*0.980*						
e. Package Error (Box d − Box a =)	*− 0.015*	*− 0.025*	*0*	*+0.030*	*− 0.070*	*− 0.020*						

Package #	Column 1 Labeled Net Weight (random pack only)	Package Errors −	Package Errors +	Column 4. MAV
1				
2				
3				
4				
5				
6				
7				
8				
9				
10				
11				
12				
Totals		f.	g.	

9. Total Error (add row e or Box f + g): *− 0 100*	10. Number of unreasonable minus (−) errors (compare each package error with the MAV in Column 4): *1*	11. Is Box 10 greater than Box 8? ☐ Yes, lot <u>fails</u> ☐ No, go to Box 12	12. Avg. error (Box 9 ÷ Box 6 =): *− 0 016*

13. Does Box 12 = Zero (0) or Plus (+)? ☐ Yes, lot <u>passes</u>, go to Box 18 ☐ No, go to Box 14	14. Compute Sample Standard Deviation:	15. Sample Correction Factor:	16. Compute Sample Error Limit (Box 14 x Box 15 =)

17. Disregarding the signs, is Box 12 larger than Box 16? ☐ Yes, lot <u>fails</u>, go to Box 18 ☐ No, lot <u>passes</u>, go to Box 18	18. Disposition of Inspection Lot ☐ Approved ☐ Rejected	19. Economic Impact: (Box 12 x Box 7 x Box 5 =) − 0.016 x $6.99 x 6 = $0.67

Comments: *Product found to contain less than the stated net contents* *Failed due to MAV*	Official's Signature: Acknowledgement of Report:

Appendix D. AOSA Rules for Testing Seeds

AOSA Rules for Testing Seeds – Section 2: Preparation of Working Samples

Volume 1. Principles and Procedures

(Provided by the Association of Official Seed Analyst)

SECTION 2: PREPARATION OF WORKING SAMPLES

The laboratory analysis for law enforcement, labeling, and general information as to seed quality, should determine the following for the sample analyzed: (1) the purity composition, (2) the rate of occurrence of noxious-weed seeds per unit weight, and (3) the percentage germination of the pure seed under consideration. Additional information, such as, seed count, detection of seed treatment, bulk examination for contaminants, tetrazolium viability, detection of fungal endophtyes, and seed moisture content may be determined using approved procedures.

2.1   Definitions

(1) **Seed unit:** the structure usually regarded as a seed in planting practices and in commercial channels. Refer to section 3.2 e for pure seed unit descriptions.

(2) **Working samples:**

(a) **Purity working sample:** the sub-sample taken from the submitted sample on which the purity analysis is performed.

2.2   Obtaining the working sample

The working sample on which the actual analysis is performed shall be taken from the submitted sample in such a manner that it will be representative. A suitable type of mechanical divider (conical, centrifugal, riffle, etc.) should be used. To avoid damage when dividing large-seeded crop kinds such as beans, peas, etc., prevent the seeds from falling great distances onto hard surfaces.

a. **Mechanical dividers.** – This method is suitable for most kinds of seeds. The apparatus divides a sample into two approximately equal parts. The submitted sample is mixed by passing it through the divider, recombining the two parts and passing the whole sample through a second time and similarly a third time. After mixing, the sample shall be reduced by passing the seed through the divider repeatedly, removing half the sample on each occasion. This process of successive halving is continued until a working sample of approximately, but not less than the minimum weight(s) stated in Table 2A is obtained.

Use of compressed air or a vacuum is highly recommended for cleaning mechanical dividers.

(1) **Centrifugal divider (Gamet type)**: This divider is suitable for all kinds of seed though it is not recommended for oilseeds (such as rapeseed, canola, mustards, flax) and kinds susceptible to damage (such as peas, soybeans, etc) and the extremely chaffy types.

The divider makes use of centrifugal force to mix and scatter seeds over the dividing surface. The seed flows downward through a hopper onto a shallow rubber cup or spinner. Upon rotation of the spinner by an electric motor the seeds are thrown out by centrifugal force and fall downward. The circle or area where the seeds fall is equally divided into two parts by a stationary baffle so that approximately half the seeds fall in one spout and half in the other spout. The centrifugal divider tends to give variable results when not carefully operated, and therefore the following procedure must be used:

(a) Preparation of the apparatus:
   (i)   Level the divider using the adjustable feet.
   (ii) Check the divider and four containers for cleanliness. Note that seeds can be trapped under the spinner and become a source of contamination.

(b) Sample mixing:
   (i)   Place a container under each spout.
   (ii) Feed the whole sample into the hopper; when filling the hopper, the seed must always be poured centrally.
   (iii) After the sample has been poured into the hopper, the spinner is operated and the seed passes into the two containers. Turn off spinner.
   (iv) Full containers are replaced by empty containers. The contents of the two full containers are fed centrally into the hopper together, the seed being allowed to blend as it flows in. The spinner is operated.
   (v)   The sample mixing procedure is repeated at least once more.

(c) Sample reduction:
   (i)   Full containers are replaced by empty containers. The contents of one full container are set aside and the contents of the other container are fed into the hopper. The spinner is operated.
   (ii) The successive halving process is continued until the working sample(s) of not less than the minimum weight(s) required stated in Table 2A are obtained.
   (iii) Ensure that the divider and containers are clean after each mixing operation.

(2) **Soil/Riffle divider**: This divider is suitable for most kinds of seed. For round-seeded kinds such as *Brassica* species, the collection containers should be covered to prevent the seeds from bouncing out.

This divider consists of a hopper with attached channels or ducts, a frame to hold the hopper, four collection containers and a pouring pan. Ducts or channels lead from the hopper to the collection containers, alternate ones leading to opposite sides. Riffle dividers are available in different sizes for different sizes of seed. The width and number of channels and spaces are important. The minimum width of the channels must be at least two times the largest diameter of the seed or any possible contaminants being mixed.

This apparatus, similar to the centrifugal divider, divides the sample into approximately equal parts.

   (a) Preparation of the apparatus:
      (i)   Place the riffle divider on a firm, level clean surface. Ensure the divider is level.
      (ii)   Ensure that the divider and the four sample collection containers are clean. Check all channels, joints and seams of the divider and collection containers to ensure there are no seeds or other plant matter present before each use.
      (iii) Two clean empty collection containers shall be placed under the channels to receive the mixed seed.

   (b) Sample mixing:
      (i)   Pour the whole sample into the divider by running the seed backwards and forwards along the edge of the divider so that all the channels and spaces of the divider receive an equal amount of seed.
      (ii)   The two full containers shall be replaced with two clean empty containers.
      (iii) The contents of one full container shall be poured into the divider by holding the long edge of the pan against the long edge of the riffle hopper and then rotating the bottom up so that the seeds pour across all channels at the same time, followed by the other full container using the same procedure.
      (iv) This process of mixing the entire submitted sample shall be repeated at least one more time before successive halving begins.

   (c) Sample reduction:
      (i)   The contents of one full container are set aside. Empty containers are placed under each channel, and the contents of the other container is poured into the hopper by holding the long edge of the pan against the long edge of the riffle hopper and then rotating the bottom up so that the seeds pour across all channels at the same time.
      (ii)   The successive halving process is continued until the working sample(s) of not less than the minimum weight(s) required stated in Table 2A are obtained.
      (iii) Ensure that the divider and collection containers are clean after each mixing operation. Check all channels of the divider, the joints and seams.

(3) **Boerner divider:** This divider is suitable for most kinds of seed, including chaffy species, peas, beans, soybeans, etc.

This divider consists of a hopper, a cone, and a series of baffles which direct the seed into two spouts. The baffles are arranged in a circle at the top and form equal width alternate channels and spaces. The channels lead to one spout, the spaces to the other. The width and number of channels and spaces are important. Five channels and spaces should be regarded as a minimum. The more channels the better but the minimum width of the channels must be at least two times the largest diameter of the seed or any possible contaminants being mixed.

   (a) Preparation of the apparatus: Ensure that the divider and the two sample collecting pans are clean.
   (b) Sample mixing:
      (i)   Place a collecting pan under each spout.
      (ii)   Close the valve at the bottom of the divider.
      (iii) Pour the seed centrally into the hopper.

(iv) Quickly open the valve. Gravity will distribute the seed evenly through the channels and spaces.

(v) To mix the seed, repeat the steps at least twice for free flowing seed and three times for chaffy grasses.

(c) Sample reduction: The contents of one full collection pan are set aside. Repeat steps in 2 "sampling mixing". To improve the randomness of reduction, choose collection pans from alternate sides for the successive halving process. The successive halving process is continued until the working sample(s) of not less than the minimum weight(s) required stated in Table 2A are obtained.

b. **Non-mechanical methods.**

(2) **Hand-halving method:** This method can be used when a proper mechanical divider is not available.

Procedure:
(a) Seed is poured evenly onto a clean smooth surface.
(b) The sample shall be thoroughly mixed using a flat-edged spatula and placed into a pile.
(c) The pile shall be divided in half using a straight edge or ruler.
(d) Each half portion is divided in half.
(e) Each of the portions is divided into half again. There are now eight portions.
(f) Arrange the eight portions into two rows of four.
(g) Alternate portions should be combined to obtain two halves e.g. combine the first portion from row 1 with the second portion from row 2. Remove the remaining four portions.
(h) Repeat steps (a) to (g) until sufficient portions of seed are taken to constitute a working sample(s) of not less than the minimum weight(s) required stated in Table 2A are obtained.

2.3 Size of working samples.

a. **Weighing the working sample.** – The weight of the working sample shall be determined to the number of decimal places indicated below:

Weight of Working Sample in Grams	Number of Decimal Places
Less than 1.000	4
1.000 to 9.999	3
10.00 to 99.99	2
100.0 to 999.9	1
1000 or more	0

AOSA Rules for Testing Seeds – Section 12: Mechanical Seed Count

Volume 1. Principles and Procedures

(Provided by the Association of Official Seed Analyst)

SECTION 12: MECHANICAL SEED COUNT

The following method shall be employed when using a mechanical seed counter to determine the number of seeds contained in a sample of soybean (*Glycine max*), corn (*Zea mays*), wheat (*Triticum aestivum*) and field bean (*Phaseolus vulgaris*).

12.1   Samples.

Samples for testing shall be of at least 500 grams for soybean, corn and field beans and 100 grams for wheat and received in moisture proof containers. Samples shall be retained in moisture proof containers until the weight of the sample prepared for purity analysis is recorded.

12.2   Seed counter calibration.

The seed counter shall be calibrated daily prior to use.

(a)   Prepare a calibration sample by counting 10 sets of 100 seeds. Visually examine each set to insure that it contains whole seeds. Combine the 10 sets of seeds to make a 1,000 seed calibration sample. The seeds of the calibration sample should be approximately the same size and shape as the seeds in a sample being tested. If the seeds in a sample being tested are noticeably different in size or shape from those in the calibration sample, prepare another calibration sample with seeds of the appropriate size and shape. Periodically re-examine the calibration samples to insure that no seeds have been lost or added.

(b)   Carefully pour the 1,000 seed calibration sample into the seed counter. Start the counter and run it until all the seeds have been counted. The seeds should not touch as they run through the counter. Record the number of seeds as displayed on the counter read out. The seed count should not vary more than ±2 seeds from 1,000. If the count is not within this tolerance, clean the mirrors, adjust the feed rate and/or reading sensitivity. Rerun the calibration sample until it is within the ±2 seed tolerance. If the seed counter continues to fail the calibration procedure and the calibration sample has been checked to ensure that it contains 1,000 seeds, do not use the counter until it has been repaired.

12.3   Sample preparation.

Immediately after opening the moisture proof container, mix and divide the submitted sample, in accordance with section 2.2, to obtain a sample for purity analysis and record the weight of this sample in grams to the appropriate number of decimal places (refer to section 2.3 a). Conduct the purity analysis to obtain pure seed for the seed count test.

12.4   Conducting the test.

After the seed counter has been calibrated, test the pure seed portion from the purity test and record the number of seeds in the sample.

12.5   Calculation of results.

Calculate the number of seeds per pound to the nearest whole number using the following formula:

$$Number\ of\ seeds\ per\ pound = \frac{453.6\ g/lb \times no.\ of\ seeds\ counted}{weight\ (g)\ of\ sample\ analyzed\ for\ purity}$$

12.6   Tolerances for results from different laboratories.

Multiply the labeled seed count or first seed count test result by four percent for soybean samples, two percent for corn (round, flat or plateless) samples, five percent for field bean samples and three percent for wheat samples. Express the tolerance (the number of seeds) to the nearest whole number. Consider the results of two tests in tolerance if the difference, expressed as the number of seeds, is equal to or less than the tolerance.

Example*:*

Kind of seed: Corn
Label claim (1st test):   2275 seed/lb.

Lab Test (2nd test):   Purity working weight = 500.3 g
Seed count of pure seed = 2479 seeds

$$Number\ of\ seeds\ per\ pound = \frac{453.6\ g/lb \times 2479\ seeds}{500.3\ g} = 2247.6\ seeds/lb$$

Rounded to the nearest whole number = 2248 seeds/lb

Calculate tolerance value for corn:

multiply label claim by 2%
2275 seeds/lb × 0.02 = 45.5 seeds/lb;
rounded to the nearest whole number = 46 seeds/lb

Determine the difference between label claim and lab test:

2275 seeds/lb – 2248 seeds/lb = 27 seeds/lb

The difference between the lab test (2nd test) and the label claim (1st test) is less than the tolerance
(27 < 46); therefore, the two results are in tolerance.

Appendix E. General Tables of Units of Measurement

These tables have been prepared for the benefit of those requiring tables of units for occasional ready reference. In Section 4. Tables of Units of Measurement of this Appendix, the tables are carried out to a large number of decimal places and exact values are indicated by underlining. In most of the other tables, only a limited number of decimal places are given, therefore making the tables better adapted to the average user.

Section 1. Tables of Metric Units of Measurement

In the metric system of measurement, designations of multiples and subdivisions of any unit may be arrived at by combining with the name of the unit the prefixes deka, hecto, and kilo meaning, respectively, 10, 100, and 1000, and deci, centi, and milli, meaning, respectively, one-tenth, one-hundredth, and one-thousandth. In some of the following metric tables, some such multiples and subdivisions have not been included for the reason that these have little, if any currency in actual usage.

In certain cases, particularly in scientific usage, it becomes convenient to provide for multiples larger than 1000 and for subdivisions smaller than one-thousandth. Accordingly, the following prefixes have been introduced and these are now generally recognized:

yotta,	(Y)	meaning 10^{24}		deci,	(d),	meaning 10^{-1}
zetta,	(Z),	meaning 10^{21}		centi,	(c),	meaning 10^{-2}
exa,	(E),	meaning 10^{18}		milli,	(m),	meaning 10^{-3}
peta,	(P),	meaning 10^{15}		micro,	(μ),	meaning 10^{-6}
tera,	(T),	meaning 10^{12}		nano,	(n),	meaning 10^{-9}
giga,	(G),	meaning 10^{9}		pico,	(p),	meaning 10^{-12}
mega,	(M),	meaning 10^{6}		femto,	(f),	meaning 10^{-15}
kilo,	(k),	meaning 10^{3}		atto,	(a),	meaning 10^{-18}
hecto,	(h),	meaning 10^{2}		zepto,	(z),	meaning 10^{-21}
deka,	(da),	meaning 10^{1}		yocto,	(y),	meaning 10^{-24}

Thus a kilometer is 1000 meters and a millimeter is 0.001 meter.

Units of Length

10 millimeters (mm)	= 1 centimeter (cm)
10 centimeters	= 1 decimeter (dm) = 100 millimeters
10 decimeters	= 1 meter (m) = 1000 millimeters
10 meters	= 1 dekameter (dam)
10 dekameters	= 1 hectometer (hm) = 100 meters
10 hectometers	= 1 kilometer (km) = 1000 meters

Units of Area

100 square millimeters (mm^2)	= 1 square centimeter (cm^2)
100 square centimeters	= 1 square decimeter (dm^2)
100 square decimeters	= 1 square meter (m^2)
100 square meters	= 1 square dekameter (dam^2) = 1 are
100 square dekameters	= 1 square hectometer (hm^2) = 1 hectare (ha)
100 square hectometers	= 1 square kilometer (km^2)

Units of Liquid Volume

10 milliliters (mL)	= 1 centiliter (cL)
10 centiliters	= 1 deciliter (dL) = 100 milliliters
10 deciliters	= 1 liter[1] = 1000 milliliters
10 liters	= 1 dekaliter (daL)
10 dekaliters	= 1 hectoliter (hL) = 100 liters
10 hectoliters	= 1 kiloliter (kL) = 1000 liters

Units of Volume

1000 cubic millimeters (mm^3)	= 1 cubic centimeter (cm^3)
1000 cubic centimeters	= 1 cubic decimeter (dm^3)
	= 1 000 000 cubic millimeters
1000 cubic decimeters	= 1 cubic meter (m^3)
	= 1 000 000 cubic centimeters
	= 1 000 000 000 cubic millimeters

Units of Mass

10 milligrams (mg)	= 1 centigram (cg)
10 centigrams	= 1 decigram (dg) = 100 milligrams
10 decigrams	= 1 gram (g) = 1000 milligrams
10 grams	= 1 dekagram (dag)
10 dekagrams	= 1 hectogram (hg) = 100 grams
10 hectograms	= 1 kilogram (kg) = 1000 grams
1000 kilograms	= 1 megagram (Mg) or 1 metric ton(t)

Section 2. Tables of U.S. Units of Measurement[2]

In these tables where <u>foot</u> or <u>mile</u> is underlined, it is survey foot or U.S. statute mile rather than international foot or mile that is meant.

---

[1] By action of the 12th General Conference on Weights and Measures (1964), the liter is a special name for the cubic decimeter.

[2] This section lists units of measurement that have traditionally been used in the United States. In keeping with the Omnibus Trade and Competitiveness Act of 1988, the ultimate objective is to make the International System of Units the primary measurement system used in the United States.

Units of Length

12 inches (in)	= 1 foot (ft)
3 feet	= 1 yard (yd)
16½ feet	= 1 rod (rd), pole, or perch
40 rods	= 1 furlong (fur) = 660 feet
8 furlongs	= 1 U.S. statute mile (mi) = 5280 feet
1852 meters (m)	= 6076.115 49 feet (approximately)
	= 1 international nautical mile

Units of Area3

144 square inches (in^2)	= 1 square foot (ft^2)
9 square feet	= 1 square yard (yd^2)
	= 1296 square inches
272¼ square feet	= 1 square rod (rd^2)
160 square rods	= 1 acre = 43 560 square feet
640 acres	= 1 square mile (mi^2)
1 mile square	= 1 section of land
6 miles square	= 1 township
	= 36 sections = 36 square miles

Units of Volume[3]

1728 cubic inches (in^3)	= 1 cubic foot (ft^3)
27 cubic feet	= 1 cubic yard (yd^3)

Gunter's or Surveyors Chain Units of Measurement

0.66 foot (ft)	= 1 link (li)
100 links	= 1 chain (ch)
	= 4 rods = 66 feet
80 chains	= 1 U.S. statute mile (mi)
	= 320 rods = 5280 feet

Units of Liquid Volume[4]

4 gills (gi)	= 1 pint (pt) = 28.875 cubic inches (in^3)
2 pints	= 1 quart (qt) = 57.75 cubic inches
4 quarts	= 1 gallon (gal) = 231 cubic inches
	= 8 pints = 32 gills

Apothecaries Units of Liquid Volume

---

[3] Squares and cubes of customary but not of metric units are sometimes expressed by the use of abbreviations rather than symbols. For example, sq ft means square foot, and cu ft means cubic foot.

[4] When necessary to distinguish the liquid pint or quart from the dry pint or quart, the word "liquid" or the abbreviation "liq" should be used in combination with the name or abbreviation of the liquid unit.

60 minims	= 1 fluid dram (fl dr or f 3)
	= 0.225 6 cubic inch (in^3)
8 fluid drams	= 1 fluid ounce (fl oz or f **3**)
	= 1.804 7 cubic inches
16 fluid ounces	= 1 pint (pt)
	= 28.875 cubic inches
	= 128 fluid drams
2 pints	= 1 quart (qt) = 57.75 cubic inches
	= 32 fluid ounces = 256 fluid drams
4 quarts	= 1 gallon (gal) = 231 cubic inches
	= 128 fluid ounces = 1024 fluid drams

Units of Dry Volume[5]

2 pints (pt)	= 1 quart (qt) = 67.200 6 cubic inches (in^3)
8 quarts	= 1 peck (pk) = 537.605 cubic inches
	= 16 pints
4 pecks	= 1 bushel (bu) = 2150.42 cubic inches
	= 32 quarts

Avoirdupois Units of Mass[6]

(The "grain" is the same in avoirdupois, troy, and apothecaries units of mass.)

$27^{11}/_{32}$ grains (gr)	= 1 dram (dr)
16 drams	= 1 ounce (oz)
	= 437½ grains
16 ounces	= 1 pound (lb)
	= 256 drams
	= 7000 grains
100 pounds	= 1 hundredweight (cwt)
20 hundredweights	= 1 ton (t)
	= 2000 pounds[7]

In "gross" or "long" measure, the following values are recognized:

112 pounds (lb)	= 1 gross or long hundredweight (cwt)[7]

---

[5] When necessary to distinguish <u>dry</u> pint or quart from the <u>liquid</u> pint or quart, the word "dry" should be used in combination with the name or abbreviation of the dry unit.

[6] When necessary to distinguish the <u>avoirdupois</u> dram from the <u>apothecaries</u> dram, or to distinguish the <u>avoirdupois</u> dram or ounce from the <u>fluid</u> dram or ounce, or to distinguish the avoirdupois ounce or pound from the <u>troy</u> or <u>apothecaries</u> ounce or pound, the word "avoirdupois" or the abbreviation "avdp" should be used in combination with the name or abbreviation of the <u>avoirdupois</u> unit.

[7] When the terms "hundredweight" and "ton" are used unmodified, they are commonly understood to mean the 100-pound hundredweight and the 2000-pound ton, respectively; these units may be designated "net" or "short" when necessary to distinguish them from the corresponding units in <u>gross</u> or <u>long</u> measure.

| 20 gross or long hundredweights | = 1 gross or long ton |
| | = 2240 pounds[7] |

Troy Units of Mass

(The "grain" is the same in avoirdupois, troy, and apothecaries units of mass.)

24 grains (gr)	= 1 pennyweight (dwt)
20 pennyweights	= 1 ounce troy (oz t) = 480 grains
12 ounces troy	= 1 pound troy (lb t)
	= 240 pennyweights = 5760 grains

Apothecaries Units of Mass

(The "grain" is the same in avoirdupois, troy, and apothecaries units of mass.)

20 grains (gr)	= 1 scruple (s ap or Э)
3 scruples	= 1 dram apothecaries (dr ap or 3)
	= 60 grains
8 drams apothecaries	= 1 ounce apothecaries (oz ap or ℥)
	= 24 scruples = 480 grains
12 ounces apothecaries	= 1 pound apothecaries (lb ap)
	= 96 drams apothecaries
	= 288 scruples = 5760 grains

Section 3. Notes on British Units of Measurement

In Great Britain, the yard, the avoirdupois pound, the troy pound, and the apothecaries pound are identical with the units of the same names used in the United States. The tables of British linear measure, troy mass, and apothecaries mass are the same as the corresponding United States tables, except for the British spelling "drachm" in the table of apothecaries mass. The table of British avoirdupois mass is the same as the United States table up to 1 pound; above that point the table reads:

14 pounds	= 1 stone
2 stones	= 1 quarter = 28 pounds
4 quarters	= 1 hundredweight= 112 pounds
20 hundredweight	= 1 ton = 2240 pounds

The present British gallon and bushel – known as the "Imperial gallon" and "Imperial bushel" – are, respectively, about 20 % and 3 % larger than the United States gallon and bushel. The Imperial gallon is defined as the volume of 10 avoirdupois pounds of water under specified conditions, and the Imperial bushel is defined as 8 Imperial gallons. Also, the subdivision of the Imperial gallon as presented in the table of British apothecaries fluid measure differs in two important respects from the corresponding United States subdivision, in that the Imperial gallon is divided into 160 fluid ounces (whereas the United States gallon is divided into 128 fluid ounces), and a "fluid scruple" is included. The full table of British measures of capacity (which are used alike for liquid and for dry commodities) is as follows:

4 gills	= 1 pint
2 pints	= 1 quart
4 quarts	= 1 gallon
2 gallons	= 1 peck

8 gallons (4 pecks)	= 1 bushel
8 bushels	= 1 quarter

The full table of British apothecaries measure is as follows:

20 minims	= 1 fluid scruple
3 fluid scruples	= 1 fluid drachm
	= 60 minims
8 fluid drachms	= 1 fluid ounce
20 fluid ounces	= 1 pint
8 pints	= 1 gallon (160 fluid ounces)

Section 4. Tables of Units of Measurement

Units of Length – International Measure[8] (all underlined figures are exact)						
Units	**Inches**	**Feet**	**Yards**	**Miles**	**Centimeters**	**Meters**
1 inch =	1	0.083 333 33	0.027 777 78	0.000 015 782 83	2.54	0.025 4
1 foot =	12	1	0.333 333 3	0.000 189 393 9	30.48	0.304 8
1 yard =	36	3	1	0.000 568 181 8	91.44	0.914 4
1 mile =	63 360	5 280	1 760	1	160 934.4	1609.344
1 centimeter =	0.393 700 8	0.032 808 40	0.010 936 13	0.000 006 213 712	1	0.01
1 meter =	39.370 08	3.280 840	1.093 613	0.000 621 371 2	100	1

Units of Length – Survey Measure[8]

(all underlined figures are exact)

Units	**Links**	**Feet**	**Rods**	**Chains**	**Miles**	**Meters**
1 link =	1	0.66	0.04	0.01	0.000 125	0.201 168 4
1 foot =	1.515 152	1	0.060 606 06	0.015 151 52	0.000 189 393 9	0.304 800 6
1 rod =	25	16.5	1	0.25	0.003 125	5.029 210
1 chain =	100	66	4	1	0.0125	20.116 84
1 mile =	8 000	5 280	320	80	1	1609.347
1 meter =	4.970 960	3.280 833	0.198 838 4	0.049 709 60	0.000 621 369 9	1

---

[8] One international foot = 0.999 998 survey foot (exactly)

    One international mile = 0.999 998 survey mile (exactly)

Units of Area – International Measure[9]

(all underlined figures are exact)

Units		Square Inches	Square Feet	Square Yards
1 square inch	=	1	0.006 944 444	0.000 771 604 9
1 square foot	=	144	1	0.111 111 1
1 square yard	=	1296	9	1
1 square mile	=	4 014 489 600	27 878 400	3 097 600
1 square centimeter	=	0.155 000 3	0.001 076 391	0.000 119 599 0
1 square meter	=	1550.003	10.763 91	1.195 990

Note: 1 survey foot $= {}^{1200}/3937$ meter (exactly)
1 international foot $= 12 \times 0.0254$ meter (exactly)
1 international foot $= 0.0254 \times 39.37$ survey foot (exactly)

Units		Square Miles	Square Centimeters	Square Meters
1 square inch	=	0.000 000 000 249 097 7	6.451 6	0.000 645 16
1 square foot	=	0.000 000 035 870 06	929.030 4	0.092 903 04
1 square yard	=	0.000 000 322 830 6	8361.273 6	0.836 127 36
1 square mile	=	1	25 899 881 103.36	2 589 988.110 336
1 square centimeter	=	0.000 000 000 038 610 22	1	0.0001
1 square meter	=	0.000 000 386 102 2	10 000	1

Units of Area – Survey Measure[9]

(all underlined figures are exact)

Units		Square Feet	Square Rods	Square Chains	Acres
1 square foot	=	1	0.003 673 095	0.000 229 568 4	0.000 022 956 84
1 square rod	=	272.25	1	0.062 5	0.006 25
1 square chain	=	4356	16	1	0.1
1 acre	=	43 560	160	10	1
1 square mile	=	27 878 400	102 400	6400	640
1 square meter	=	10.763 87	0.039 536 70	0.002 471 044	0.000 247 104 4
1 hectare	=	107 638.7	395.367 0	24.710 44	2.471 044

---

[9] One square survey foot $= 1.000\ 004$ square international feet
One square survey mile $= 1.000\ 004$ square international miles
[9] One square survey foot $= 1.000\ 004$ square international feet
One square survey mile $= 1.000\ 004$ square international miles

Units		Square Miles	Square Meters	Hectares
1 square foot	=	0.000 000 035 870 06	0.092 903 41	0.000 009 290 341
1 square rod	=	0.000 009 765 625	25.292 95	0.002 529 295
1 square chain	=	0.000 156 25	404.687 3	0.040 468 73
1 acre	=	0.001 562 5	4 046.873	0.404 687 3
1 square mile	=	1	2 589 998	258.999 8
1 square meter	=	0.000 000 386 100 6	1	0.000 1
1 hectare	=	0.003 861 006	10 000	1

Units of Volume

(all underlined figures are exact)

Units		Cubic Inches	Cubic Feet	Cubic Yards
1 cubic inch	=	1	0.000 578 703 7	0.000 021 433 47
1 cubic foot	=	1728	1	0.037 037 04
1 cubic yard	=	46 656	27	1
1 cubic centimeter	=	0.061 023 74	0.000 035 314 67	0.000 001 307 951
1 cubic decimeter	=	61.023 74	0.035 314 67	0.001 307 951
1 cubic meter	=	61 023.74	35.314 67	1.307 951

Units		Milliliters (Cubic Centimeters)	Liters (Cubic Decimeters)	Cubic Meters
1 cubic inch	=	16.387 064	0.016 387 064	0.000 016 387 064
1 cubic foot	=	28 316.846 592	28.316 846 592	0.028 316 846 592
1 cubic yard	=	764 554 857 984	764.554 857 984	0.764 554 857 984
1 cubic centimeter	=	1	0.001	0.000 001
1 cubic decimeter	=	1000	1	0.001
1 cubic meter	=	1 000 000	1000	1

Units of Capacity or Volume – Dry Volume Measure

(all underlined figures are exact)

Units		Dry Pints	Dry Quarts	Pecks	Bushels
1 dry pint	=	1	0.5	0.062 5	0.015 625
1 dry quart	=	2	1	0.125	0.031 25
1 peck	=	16	8	1	0.25
1 bushel	=	64	32	4	1
1 cubic inch	=	0.029 761 6	0.014 880 8	0.001 860 10	0.000 465 025
1 cubic foot	=	51.428 09	25.714 05	3.214 256	0.803 563 95
1 liter	=	1.816 166	0.908 083 0	0.113 510 4	0.028 377 59
1 cubic meter	=	1 816.166	908.083 0	113.510 4	28.377 59

Units		Cubic Inches	Cubic Feet	Liters	Cubic Meters
1 dry pint	=	33.600 312 5	0.019 444 63	0.550 610 5	0.000 550 610 5
1 dry quart	=	67.200 625	0.038 889 25	1.101 221	0.001 101 221
1 peck	=	537.605	0.311 114	8.809 768	0.008 809 768
1 bushel	=	2 150.42	1.244 456	35.239 07	0.035 239 07
1 cubic inch	=	1	0.000 578 703 7	0.016 387 06	0.000 016 387 06
1 cubic foot	=	1728	1	28.316 85	0.028 316 85
1 liter	=	61.023 74	0.035 314 67	1	0.001
1 cubic meter	=	61 023.74	35.314 67	1000	1

Units of Capacity or Volume – Liquid Volume Measure

(All underlined figures are exact)

Units		Minims	Fluid Drams	Fluid Ounces	Gills
1 minim	=	1	0.016 666 67	0.002 083 333	0.000 520 833 3
1 fluid dram	=	60	1	0.125	0.031 25
1 fluid ounce	=	480	8	1	0.25
1 gill	=	1 920	32	4	1
1 liquid pint	=	7 680	128	16	4
1 liquid quart	=	15 360	256	32	8
1 gallon	=	61 440	1024	128	32
1 cubic inch	=	265.974 0	4.432 900	0.554 112 6	0.138 528 1
1 cubic foot	=	459 603.1	7660.052	957.506 5	239.376 6
1 milliliter	=	16.230 73	0.270 512 2	0.033 814 02	0.008 453 506
1 liter	=	16 230.73	270.512 2	33.814 02	8.453 506

Units		Liquid Pints	Liquid Quarts	Gallons	Cubic Inches
1 minim	=	0.000 130 208 3	0.000 065 104 17	0.000 016 276 04	0.003 759 766
1 fluid dram	=	0.007 812 5	0.003 906 25	0.000 976 562 5	0.225 585 94
1 fluid ounce	=	0.062 5	0.031 25	0.007 812 5	1.804 687 5
1 gill	=	0.25	0.125	0.031 25	7.218 75
1 liquid pint	=	1	0.5	0.125	28.875
1 liquid quart	=	2	1	0.25	57.75
1 gallon	=	8	4	1	231
1 cubic inch	=	0.034 632 03	0.017 316 02	0.004 329 004	1
1 cubic foot	=	59.844 16	29.922 08	7.480 519	1728
1 milliliter	=	0.002 113 376	0.001 056 688	0.000 264 172 1	0.061 023 74
1 liter	=	2.113 376	1.056 688	0.264 172 1	61.023 74

152

Units		Cubic Feet	Milliliters	Liters
1 minim	=	0.000 002 175 790	0.061 611 52	0.000 061 611 52
1 fluid dram	=	0.000 130 547 4	3.696 691	0.003 696 691
1 fluid ounce	=	0.001 044 379	29.573 53	0.029 573 53
1 gill	=	0.004 177 517	118.294 1	0.118 294 1
1 liquid pint	=	0.016 710 07	473.176 5	0.473 176 5
1 liquid quart	=	0.033 420 14	946.352 9	0.946 352 9
1 gallon	=	0.133 680 6	3785.412	3.785 412
1 cubic inch	=	0.000 578 703 7	16.387 06	0.016 387 06
1 cubic foot	=	1	28 316.85	28.316 85
1 milliliter	=	0.000 035 314 67	1	0.001
1 liter	=	0.035 314 67	1000	1

Units of Mass Not Less Than Avoirdupois Ounces

(all underlined figures are exact)

Units		Avoirdupois Ounces	Avoirdupois Pounds	Short Hundred-weights	Short Tons
1 avoirdupois ounce	=	1	0.0625	0.000 625	0.000 031 25
1 avoirdupois pound	=	16	1	0.01	0.000 5
1 short hundredweight	=	1 600	100	1	0.05
1 short ton	=	32 000	2000	20	1
1 long ton	=	35 840	2240	22.4	1.12
1 kilogram	=	35.273 96	2.204 623	0.022 046 23	0.001 102 311
1 metric ton	=	35 273.96	2204.623	22.046 23	1.102 311

Units		Long Tons	Kilograms	Metric Tons
1 avoirdupois ounce	=	0.000 027 901 79	0.028 349 523 125	0.000 028 349 523 125
1 avoirdupois pound	=	0.000 446 428 6	0.453 592 37	0.000 453 592 37
1 short hundredweight	=	0.044 642 86	45.359 237	0.045 359 237
1 short ton	=	0.892 857 1	907.184 74	0.907 184 74
1 long ton	=	1	1016.046 908 8	1.016 046 908 8
1 kilogram	=	0.000 984 206 5	1	0.001
1 metric ton	=	0.984 206 5	1000	1

Units of Mass Not Greater Than Pounds and Kilograms

(all underlined figures are exact)

Units		Grains	Apothecaries Scruples	Pennyweights	Avoirdupois Drams
1 grain	=	1	0.05	0.041 666 67	0.036 571 43
1 apoth. scruple	=	20	1	0.833 333 3	0.731 428 6
1 pennyweight	=	24	1.2	1	0.877 714 3
1 avdp. dram	=	27.343 75	1.367 187 5	1.139 323	1
1 apoth. dram	=	60	3	2.5	2.194 286
1 avdp. ounce	=	437.5	21.875	18.229 17	16
1 apoth. or troy oz.	=	480	24	20	17.554 29
1 apoth. or troy pound	=	5760	288	240	210.651 4
1 avdp. pound	=	7000	350	291.666 7	256
1 milligram	=	0.015 432 36	0.000 771 617 9	0.000 643 014 9	0.000 564 383 4
1 gram	=	15.432 36	0.771 617 9	0.643 014 9	0.564 383 4
1 kilogram	=	15432.36	771.617 9	643.014 9	564.383 4

Units		Apothecaries Drams	Avoirdupois Ounces	Apothecaries or Troy Ounces	Apothecaries or Troy Pounds
1 grain	=	0.016 666 67	0.002 285 714	0.002 083 333	0.000 173 611 1
1 apoth. scruple	=	0.333 333 3	0.045 714 29	0.041 666 67	0.003 472 222
1 pennyweight	=	0.4	0.054 857 14	0.05	0.004 166 667
1 avdp. dram	=	0.455 729 2	0.062 5	0.56 966 15	0.004 747 179
1 apoth. dram	=	1	0.137 142 9	0.125	0.010 416 67
1 avdp. ounce	=	7.291 667	1	0.911 458 3	0.075 954 86
1 apoth. or troy ounce	=	8	1.097 143	1	0.083 333 333
1 apoth. or troy pound	=	96	13.165 71	12	1
1 avdp. pound	=	116.666 7	16	14.583 33	1.215 278
1 milligram	=	0.000 257 206 0	0.000 035 273 96	0.000 032 150 75	0.000 002 679 229
1 gram	=	0.257 206 0	0.035 273 96	0.032 150 75	0.002 679 229
1 kilogram	=	257.206 0	35.273 96	32.150 75	2.679 229

Units		Avoirdupois Pounds	Milligrams	Grams	Kilograms
1 grain	=	0.000 142 857 1	64.798 91	0.064 798 91	0.000 064 798 91
1 apoth. scruple	=	0.002 857 143	1295.978 2	1.295 978 2	0.001 295 978 2
1 pennyweight	=	0.003 428 571	1555.173 84	1.555 173 84	0.001 555 173 84
1 avdp. dram	=	0.003 906 25	1771.845 195 312 5	1.771 845 195 312 5	0.001 771 845 195 312 5
1 apoth. dram	=	0.008 571 429	3887.934 6	3.887 934 6	0.003 887 934 6
1 avdp. ounce	=	0.062 5	28 349.523 125	28.349 523 125	0.028 349 523 125
1 apoth. or troy ounce	=	0.068 571 43	31 103.476 8	31.103 476 8	0.031 103 476 8
1 apoth. or troy pound	=	0.822 857 1	373 241.721 6	373.241 721 6	0.373 241 721 6
1 avdp. pound	=	1	453 592.37 37	453.592 37	0.453 592 37
1 milligram	=	0.000 002 204 623	1	0.001	0.000 001
1 gram	=	0.002 204 623	1000	1	0.001
1 kilogram	=	2.204 623	1 000 000	1000	1

Section 5. Tables of Equivalents

In these tables it is necessary to differentiate between the "international foot" and the "survey foot." Therefore, the survey foot is underlined.

When the name of a unit is enclosed in brackets (thus, [1 hand] . . .), this indicates (1) that the unit is not in general current use in the United States, or (2) that the unit is believed to be based on "custom and usage" rather than on formal authoritative definition.

Equivalents involving decimals are, in most instances, rounded off to the third decimal place except where they are exact, in which cases these exact equivalents are so designated. The equivalents of the imprecise units "tablespoon" and "teaspoon" are rounded to the nearest milliliter.

Units of Length	
angstrom (Å)[10]	0.1 nanometer (exactly) 0.000 1 micrometer (exactly) 0.000 000 1 millimeter (exactly) 0.000 000 004 inch
1 cable's length	120 fathoms (exactly) 720 feet (exactly) 219 meters
1 centimeter (cm)	0.393 7 inch
1 chain (ch)	66 feet (exactly)

[10] The angstrom is basically defined as 10^{-10} meter.

Units of Length	
(Gunter's or surveyors)	20.116 8 meters
1 decimeter (dm)	3.937 inches
1 dekameter (dam)	32.808 feet
1 fathom	6 feet (exactly)
	1.828 8 meters
1 foot (ft)	0.304 8 meter (exactly)
1 furlong (fur)	10 chains (surveyors) (exactly)
	660 feet (exactly)
	$^1/_8$ U.S. statute mile (exactly)
	201.168 meters
[1 hand]	4 inches
1 inch (in)	2.54 centimeters (exactly)
1 kilometer (km)	0.621 mile
1 league (land)	3 U.S. statute miles (exactly)
	4.828 kilometers
1 link (li) (Gunter's or surveyors)	0.66 foot (exactly)
	0.201 168 meter
1 meter (m)	39.37 inches
	1.094 yards
1 micrometer	0.001 millimeter (exactly)
	0.000 039 37 inch
1 mil	0.001 inch (exactly)
	0.025 4 millimeter (exactly)
1 mile (mi) (U.S. statute)[11]	5280 feet survey (exactly)
	1.609 kilometers
1 mile (mi) (international)	5280 feet international (exactly)
1 mile (mi) (international nautical)[12]	1.852 kilometers (exactly)
	1.151 survey miles
1 millimeter (mm)	0.039 37 inch
	0.001 meter (exactly)
1 nanometer (nm)	0.000 000 039 37 inch
1 Point (typography)	0.013 837 inch (exactly)
	$^1/_{72}$ inch (approximately)
	0.351 millimeter
1 rod (rd), pole, or perch	16½ feet (exactly)
	5.029 2 meters
1 yard (yd)	0.914 4 meter (exactly)

---

[11] The term "statute mile" originated with Queen Elizabeth I who changed the definition of the mile from the Roman mile of 5000 feet to the statute mile of 5280 feet. The international mile and the U.S. statute mile differ by about 3 millimeters although both are defined as being equal to 5280 feet. The international mile is based on the international foot (0.3048 meter) whereas the U.S. statute mile is based on the survey foot (1200/3937 meter).

[12] The international nautical mile of 1852 meters (6076.115 49 feet) was adopted effective July 1, 1954, for use in the United States. The value formerly used in the United States was 6080.20 feet = 1 nautical (geographical or sea) mile.

Units of Area	
1 acre[13]	43 560 square <u>feet</u> (exactly) 0.405 hectare
1 are	119.599 square yards 0.025 acre
1 hectare	2.471 acres
[1 square (building)]	100 square feet
1 square centimeter (cm^2)	0.155 square inch
1 square decimeter (dm^2)	15.500 square inches
1 square foot (ft^2)	929.030 square centimeters
1 square inch (in^2)	6.451 6 square centimeters (exactly)
1 square kilometer (km^2)	247.104 acres 0.386 square mile
1 square meter (m^2)	1.196 square yards 10.764 square feet
1 square mile (mi^2)	258.999 hectares
1 square millimeter (mm^2)	0.002 square inch
1 square rod (rd^2), sq pole, or sq perch	25.293 square meters
1 square yard (yd^2)	0.836 square meter

Units of Capacity or Volume	
1 barrel (bbl), liquid	31 to 42 gallons[14]

1 barrel (bbl), standard for fruits,
  vegetables, and other dry
  commodities, except cranberries
	7056 cubic inches
	105 dry quarts
	3.281 bushels, struck measure

1 barrel (bbl), standard, cranberry	5826 cubic inches $86^{45}/_{64}$ dry quarts 2.709 bushels, struck measure
1 bushel (bu) (U.S.) struck measure	2150.42 cubic inches (exactly) 35.238 liters
[1 bushel, heaped (U.S.)]	2747.715 cubic inches 1.278 bushels, struck measure[15]
[1 bushel (bu) (British Imperial) (struck measure)]	1.032 U.S. bushels, struck measure 2219.36 cubic inches
1 cord (cd) (firewood)	128 cubic feet (exactly)
1 cubic centimeter (cm^3)	0.061 cubic inch
1 cubic decimeter (dm^3)	61.024 cubic inches
1 cubic foot (ft^3)	7.481 gallons 28.316 cubic decimeters

---

[13] The question is often asked as to the length of a side of an acre of ground. An acre is a unit of area containing 43 560 square <u>feet</u>. It is not necessarily square, or even rectangular. But, if it is square, then the length of a side is equal to $\sqrt{43560 \text{ ft}^2} = 208.710$ ft (not exact).

[14] There are a variety of "barrels" established by law or usage. For example, federal taxes on fermented liquors are based on a barrel of 31 gallons; many state laws fix the "barrel for liquids" as 31½ gallons; one state fixes a 36-gallon barrel for cistern measurement; federal law recognizes a 40-gallon barrel for "proof spirits;" by custom, 42 gallons comprise a barrel of crude oil or petroleum products for statistical purposes, and this equivalent is recognized "for liquids" by four states.

[15] Frequently recognized as 1¼ bushels, struck measure.

Units of Capacity or Volume	
1 cubic inch (in^3)	0.554 fluid ounce 4.433 fluid drams 16.387 cubic centimeters
1 cubic meter (m^3)	1.308 cubic yards
1 cubic yard (yd^3)	0.765 cubic meter
1 cup, measuring	8 fluid ounces (exactly) 237 milliliters ½ liquid pint (exactly)
1 dekaliter (daL)	2.642 gallons 1.135 pecks
1 dram, fluid (or liquid) (fl dr) or f^3 (U.S.)	$^1/_8$ fluid ounce (exactly) 0.226 cubic inch 3.697 milliliters 1.041 British fluid drachms
[1 drachm, fluid (fl dr) (British)]	0.961 U.S. fluid dram 0.217 cubic inch 3.552 milliliters
1 gallon (gal) (U.S.)	231 cubic inches (exactly) 3.785 liters 0.833 British gallon 128 U.S. fluid ounces (exactly)
[1 gallon (gal) (British Imperial)]	277.42 cubic inches 1.201 U.S. gallons 4.546 liters 160 British fluid ounces (exactly)
1 gill (gi)	7.219 cubic inches 4 fluid ounces (exactly) 0.118 liter
1 hectoliter (hL)	26.418 gallons 2.838 bushels
1 liter (1 cubic decimeter exactly)	1.057 liquid quarts 0.908 dry quart 61.025 cubic inches
1 milliliter (mL)	0.271 fluid dram 16.231 minims 0.061 cubic inch
1 ounce, fluid (or liquid) (fl oz) or $f\,^{\overline{3}}$) (U.S.)	1.805 cubic inches 29.573 milliliters 1.041 British fluid ounces
[1 ounce, fluid (fl oz) (British)]	0.961 U.S. fluid ounce 1.734 cubic inches 28.412 milliliters
1 peck (pk)	8.810 liters
1 pint (pt), dry	33.600 cubic inches 0.551 liter
1 pint (pt), liquid	28.875 cubic inches (exactly 0.473 liter
1 quart (qt), dry (U.S.)	67.201 cubic inches 1.101 liters 0.969 British quart
1 quart (qt), liquid (U.S.)	57.75 cubic inches (exactly) 0.946 liter 0.833 British quart

Units of Capacity or Volume	
[1 quart (qt) (British)]	69.354 cubic inches 1.032 U.S. dry quarts 1.201 U.S. liquid quarts
1 tablespoon, measuring	3 teaspoons (exactly) 15 milliliters 4 fluid drams ½ fluid ounce (exactly)
1 teaspoon, measuring	⅓ tablespoon (exactly) 5 milliliters 1⅓ fluid drams[16]
1 water ton (English)	270.91 U.S. gallons 224 British Imperial gallons (exactly)

Units of Mass	
1 assay ton[17] (AT)	29.167 grams
1 carat (c)	200 milligrams (exactly) 3.086 grains
1 dram apothecaries (dr ap or ʒ)	60 grains (exactly) 3.888 grams
1 dram avoirdupois (dr avdp)	$27^{11}/_{32}$ (= 27.344) grains 1.772 grams
1 gamma (γ)	1 microgram (exactly)
1 grain	64.798 91 milligrams (exactly)
1 gram (g)	15.432 grains 0.035 ounce, avoirdupois
1 hundredweight, gross or long[18] (gross cwt)	112 pounds (exactly) 50.802 kilograms
1 hundredweight, gross or short (cwt or net cwt)	100 pounds (exactly) 45.359 kilograms
1 kilogram (kg)	2.205 pounds
1 microgram (μg) [the Greek letter mu in combination with the letter g]	0.000 001 gram (exactly)
1 milligram (mg)	0.015 grain
1 ounce, avoirdupois (oz avdp)	437.5 grains (exactly) 0.911 troy or apothecaries ounce 28.350 grams

---

[16] The equivalent "1 teaspoon = 1⅓ fluid drams" has been found by the Bureau to correspond more closely with the actual capacities of "measuring" and silver teaspoons than the equivalent "1 teaspoon = 1 fluid dram," which is given by a number of dictionaries.

[17] Used in assaying. The assay ton bears the same relation to the milligram that a ton of 2000 pounds avoirdupois bears to the ounce troy; hence the mass in milligrams of precious metal obtained from one assay ton of ore gives directly the number of troy ounces to the net ton.

[18] The gross or long ton and hundredweight are used commercially in the United States to only a very limited extent, usually in restricted industrial fields. The units are the same as the British "ton" and "hundredweight."

Units of Mass	
1 ounce, troy or apothecaries (oz t or oz ap or ℥)	480 grains (exactly) 1.097 avoirdupois ounces 31.103 grams
1 pennyweight (dwt)	1.555 grams
1 point	0.01 carat 2 milligrams
1 pound, avoirdupois (lb avdp)	7000 grains (exactly) 1.215 troy or apothecaries pounds 453.592 37 grams (exactly)
1 pound, troy or apothecaries (lb t or lb ap)	5760 grains (exactly) 0.823 avoirdupois pound 373.242 grams
1 scruple (s ap or ℈)	20 grains (exactly) 1.296 grams
1 ton, gross or long[19]	2240 pounds (exactly) 1.12 net tons (exactly) 1.016 metric tons
1 ton, metric (t)	2204.623 pounds 0.984 gross ton 1.102 net tons
1 ton, net or short	2000 pounds (exactly) 0.893 gross ton 0.907 metric ton

---

[19] The gross or long ton and hundredweight are used commercially in the United States to a limited extent only, usually in restricted industrial fields. These units are the same as the British "ton" and "hundredweight."

Appendix F. Glossary

A

allowable difference. The amount, by which the actual quantity in the package may differ from the declared quantity. Pressed and blown tumblers and stemware labeled by count and capacity are assigned an allowable difference in capacity. This is also called a tolerance.

audit testing. Preliminary tests designed to quickly identify potential noncompliance units.

average. The sum of a number of individual measurement values divided by the number of values. For example, the sum of the individual weights of 12 packages divided by 12 would be the average weight of those packages.

average error. The sum of the individual "package errors" (defined) (considering their arithmetic sign) divided by the number of packages comprising the sample.

average requirement. A requirement that the average net quantity of contents of packages in a "lot" equals the net quantity of contents printed on the label.

average tare. The sum of the weights of individual package containers (or wrappers, etc.) divided by the number of containers or wrappers weighed.

B

berry baskets and boxes. Disposable containers in capacities of 1 dry quart or less for berries and small fruits. See Section 4.46. in NIST Handbook 44.

C

Category A (Category B). A set of sampling plans provided in this handbook to use in checking packages that must (except when exempted) meet the "average requirement" (defined).

chamois. A natural leather made from skins of sheep and lambs that have been oil-tanned.

combination quantity declarations. A package label that contains the count of items in the package as well as one or more of the following: weight, measure, or size.

compliance testing. Determining package conformance using specified legal requirements.

D

decision criteria. The rules for deciding whether or not a lot conforms to package requirements based on the results of checking the packages in the sample.

delivery. A quantity of identically labeled product received at one time by a buyer.

dimensionless units. The integers in terms of which the official records package errors. The dimensionless units must be multiplied by the "unit of measure" to obtain package errors in terms of weight, length, etc.

division, value of (d). The value of the scale division, expressed in units of mass, is the smallest subdivision of the scale for analog indication or the difference between two consecutively indicated or printed values for digital indication or printing. See NIST Handbook 44.

drained weight. The weight of solid or semisolid product representing the contents of a package obtained after a prescribed method for removal of the liquid has been employed.

dry measure. Rigid containers designed for general and repeated use in the volume measurement of particulate solids. See Section 4.45. Dry Measures in NIST Handbook 44.

dry pet food. All extruded dog and cat foods and baked treats packaged in Kraft paper bags and cardboard boxes that have a moisture content of 13 % or less at the time of packaging.

dry tare. See UNUSED DRY TARE.

E

error. See PACKAGE ERROR.

G

gravimetric test procedure. An analytical procedure that involves measurement by mass or weight.

gross weight. The weight of the package including contents, packing material, labels.

H

headspace. The container volume not occupied by product.

I

inch-pound units. Units based upon the yard, gallon, and the pound commonly used in the United States of America. Some of these units have the same name as similar units in the United Kingdom (British, English, or Imperial units), but they are not necessarily equal to them.

initial tare sample. The first packages (either two or five) selected from the sample to be opened for tare determination in the tare procedure. Depending upon the variability of these individual tare weights as compared with the variability of the net contents, this initial tare sample may be sufficient or more packages may be needed to determine the tare.

inspection lot. The collection of identically labeled (random packages, in some cases, are exempt from identity and labeled quantity when determining the inspection lot) packages available for inspection at one time. This collection will pass or fail as a whole based on the results of tests on a sample drawn from this collection.

L

label. Any written, printed, or graphic matter affixed to, applied to, attached to, blown into, formed, molded into, embossed on, or appearing upon or adjacent to a consumer commodity or a package containing any consumer commodity, for purposes of branding, identifying, or giving any information with respect to the commodity or to the contents of the package, except that an inspector's tag or other non-promotional matter affixed to or appearing upon a consumer commodity is not a label. See Section 2.5 in the Uniform Packaging and Labeling Regulation in NIST Handbook 130.

linear measures. Rulers and tape measures.

location of test. The place where the package will be examined. This is broadly defined as one of three general locations: (1) where the commodity was packaged, (2) a warehouse or storage location, or (3) a retail outlet.

lot. See INSPECTION LOT.

lot code. A series of identifying numbers and/or letters on the outside of a package designed to provide information such as the date and location of packaging or the expiration date.

lot size. The number of packages in the "inspection lot".

M

MAV. See MAXIMUM ALLOWABLE VARIATION

maximum allowable variation (MAV). A deficiency in the weight, measure, or count of an individual package beyond which the deficiency is considered to be an "unreasonable error". The number of packages with deficiencies that are greater than the MAV is controlled by the sampling procedure.

measure containers. Containers whose capacities are used to determine quantity. They are of two basic types: (a) retail and (b) prepackaged. Retail containers are packaged at the time of retail sale, and prepackaged containers are packaged in advance of sale. An example of a prepackaged measure container is an ice cream package.

metric or SI units. Units of the International System of Units as established in 1960 by the General Conference on Weights and Measures and interpreted or modified for the United States by the Secretary of Commerce. (See NIST Special Publication 814 – Metric System of Measurement; Interpretation of the SI for the United States and Federal Government Metric Conversion Policy)

minus or plus errors. Negative or positive deviations from the labeled quantity of the actual package quantities as measured. See PACKAGE ERROR.

moisture allowance. That variation in weight of a packaged product permitted in order to account for loss of weight due to loss of moisture during good package distribution practices. For packaged goods subject to moisture loss, when the average net weight of a sample is found between the labeled weight and the boundary of the moisture allowance, the lot is said to be in a no-decision area. Further information is required to determine lot compliance or noncompliance.

mulch. Any product or material other than peat or peat moss for sale, or sold for primary use as a horticultural, above-ground dressing for decoration, moisture control, weed control, erosion control, temperature control, or other similar purposes.

N

net quantity or net contents. That quantity of packaged product remaining after all necessary deductions for tare (defined) have been made.

nominal. A designated or theoretical size that may vary from the actual.

nominal gross weight. The sum of the nominal tare weight (defined) plus the declared or labeled weight (or other labeled quantity converted to a weight basis).

P

package error. The difference between the actual net contents of an individual package as measured and the declared net contents on the package label; minus (−) for less than the label and plus (+) for more than the label.

packaged goods. Product or commodity put up in any manner in advance of sale suitable for either wholesale or retail sale.

petroleum products. Gasoline, diesel fuel, kerosene, or any product (whether or not such a product is actually derived from naturally occurring hydro-carbon mixtures known as "petroleum") commonly used in powering, lubricating, or idling engines or other devices, or labeled as fuel to power camping stoves or lights. Sewing machine lubricant, camping fuels, and synthetic motor oil are "petroleum products" for the purposes of this regulation. The following products are not "petroleum products": brake fluid, copier machine dispersant, antifreeze, cleaning solvents, and alcohol.

plus errors. See MINUS OR PLUS ERRORS

principal display panel or panels. Part(s) of a label that are designed to be displayed, presented, shown, or examined under normal and customary conditions of display and purchase. Wherever a principal display panel appears more than once on a package, all requirements pertaining to the "principal display panel" shall pertain to all such "principal display panels." See Section 2.7 in the Uniform Packaging and Labeling Regulation in NIST Handbook 130.

production lot. The total collection of packages defined by the packager, usually consisting of those packages produced within a given unit of time and coded identically.

pycnometer. A container of known volume used to contain material for weighing so that the weight of a known volume may be determined for the material. If it is constructed, it is called a density cup.

R

random pack. The term "random package" shall be construed to mean a package that is one of a lot, shipment, or delivery of packages of the same consumer commodity with varying weights which means, packages of the same consumer commodity with no fixed pattern of weight.

random sampling. The process of selecting sample packages such that all packages under consideration have the same probability of being selected. An acceptable method of random selection is to use a table of random numbers.

range. The difference between the largest and the smallest of a set of measured values.

reasonable variation. An amount by which individual package net contents are allowed to vary from the labeled net contents. This term is found in most federal and state laws and regulations governing packaged goods. Reasonable variations from the labeled declaration are recognized for (1) unavoidable deviations in good manufacturing practice, and (2) loss or gain of moisture in good distribution practice.

rounding. The process of omitting some of the end digits of a numerical value and adjusting the last retained digit so that the resulting number is as near as possible to the original number.

S

sample. A group of packages taken from a larger collection of packages and providing information that can be used to make a decision concerning the larger collection of packages or of the package production process. A sample provides a valid basis for decision only when it is a random sample (defined).

sample correction factor. The factor as computed is the ratio of the 97.5[th] quantile of the student's t-distribution with (n-1) degrees of freedom and the square root of n where n is the sample size.

sample error limit (SEL). A statistical value computed by multiplying the sample standard deviation times the sample correction factor from Column 3 of Table 2-1. Category A – Sampling Plans for the appropriate sample size. The SEL value allows for the uncertainty between the average error of the sample and the average error of the inspection lot with an approximately 97.5 % level of confidence.

sample size (n). The number of packages in a sample.

sampling plan. A specific plan that states the number of packages to be checked and the associated decision criteria.

scale tolerance. The official value fixing the limit of allowable error for weighing equipment as defined in NIST Handbook 44.

seat. (as in "seat diameter" or "seated capacity"). The projection or shoulder near the upper rim of a cup or container that is designed to serve as the support for a lid or cover.

seated capacity. The capacity of a cup, container, or bottle, as defined by the volume contained by them when the lid or a flat disc is inserted into the lid groove that is located inside and near the upper rim of the cup, container, or bottle.

SEL. See SAMPLE ERROR LIMIT.

shipment. A quantity of identically labeled product (except for lot code) sent at one time to a single location.

slicker plate. A flat plate, usually of glass or clear plastic composition, used to determine the "level full" condition of a capacity (volumetric) measure.

standard deviation. A measure to describe the scatter of the individual package contents around the mean contents.

standard pack. That type of package in which a commodity is put up with identical labels and only in certain specific quantity sizes. Examples of goods so packed are canned, boxed, bottled and bagged foods, and over-the-counter drugs.

supplementary quantity declarations. The required quantity declaration may be supplemented by one or more declarations of weight, measure, or count, such declaration appearing other than on a principal display panel. Such supplemental statement of quantity of contents shall not include any terms qualifying a unit of weight, measure, or count that tends to exaggerate the amount of commodity contained in the package (e.g., "giant" quart, "full" gallon, "when packed," "minimum," or words of similar import). See Section 6.12 in the Uniform Packaging and Labeling Regulation in NIST Handbook 130.

T

tare sample. The packages or packaging material used to determine the average tare weight.

tare sample size. The number of packages or packaging material units used to determine the average tare weight.

tare weight. The weight of a container, wrapper, or other material that is deducted from the gross weight to obtain the net weight.

tolerance. A value fixing the limit of allowed departure from the labeled contents; usually presented as a plus (+) and minus (-) value.

U

unit of measure. An increment of weight, length, or volume so that an inspector may record package errors in terms of small integers. (The package errors are actually the integers multiplied by the unit of measure.)

unreasonable errors. Minus package errors that exceed the MAV (defined). The number of unreasonable errors permitted in a sample is specified by the sampling plan.

unused dry tare. All unused packaging materials (including glue, labels, ties, etc.) that contain or enclose a product. It includes prizes, gifts, coupons, or decorations that are not part of the product.

used dry tare. Used tare material that has been air dried, or dried in some manner to simulate the unused tare weight. It includes all packaging materials that can be separated from the packaged product, either readily (e.g., by shaking) or by washing, scraping, ambient air drying, or other techniques involving more than "normal" household recovery procedures, but not including laboratory procedures like oven drying. Labels, wire closures, staples, prizes, decorations, and such are considered tare. It is not the same as "wet tare." See also "wet tare."

V

volumetric measures. Standard measuring flasks, graduates, cylinders, for use in measuring volumes of liquids.

W

wet tare. Used packaging materials when no effort is made to reconstruct unused tare weight by drying out the absorbent portion (if any) of the tare.

THIS PAGE INTENTIONALLY LEFT BLANK

Index

References

2010 AOSA Rules for Testing Seeds, Volume 1 Section 2 and Section 12. Association of Official Seed Analyst (AOSA), Inc. 101 East State Street #214, Ithaca, NY 14850

C. Brickenkamp, S. Hasko, and M. G. Natrella, Third Edition of NIST Handbook 133 – Checking the Net Contents of Packaged Goods, 1988.

K. Butcher and T. Coleman, 4th Supplement to the Third Edition of NIST Handbook 133 – Checking the Net Contents of Packaged Goods, 1994.

T. Coleman, L. Crown and K. Dresser, Fourth Edition NIST Handbook 133 – Checking the Net Contents of Packaged Goods, 2005. Available at http://www.nist.gov/pml/wmd

T. Butcher, S. Cook, and L. Crown, Specifications, Tolerances, and Other Technical Requirements for Weighing and Measuring Devices, National Institute of Standards and Technology Handbook 44, 2010. Available at http://www.nist.gov/pml/wmd

L. Crown, D. Sefcik and L. Warfield, Uniform Laws and Regulations in the Areas of Legal Metrology and Engine Fuel Quality, National Institute of Standards and Technology Handbook 130, 2010. Available at http://www.nist.gov/pml/wmd

Compressed Gas Association, Fourth Edition – Handbook of Compressed Gases, 1999. Compressed Gas Association, 4221 Walney Road, 5th Floor, Chantilly, Virginia 20151-2923. Available at http://www.cganet.com

Compressed Gas Association - pamphlet P-1, "Safe Handling of Compressed Gases in Containers, Compressed Gas Association, 4221 Walney Road, 5th Floor, Chantilly, Virginia 20151-2923. Available at http://www.cganet.com

P. Cunniff, ed., Official Methods of Analysis of the Association of Official Analytical Chemists International, Seventeenth Edition, Association of Official Analytical Chemists, 481 North Frederick Avenue, Suite 500, Gaithersburg, Maryland 20877, 2003. Available at http://www.aoac.org

Federal Test Method Standard 311 "Leather, Methods of Sampling and Testing." (January 15, 1969). U.S. General Services Administration.

G. L. Harris, Specifications and Tolerances for Reference Standards and Field Standard Weights and Measures, 1. Specifications and Tolerances for Field Standard Weights (National Institute of Standards and Technology Class F), National Institute of Standards and Technology Handbook 105-1, 1990. Available at http://www.nist.gov/pml/wmd

G. L. Harris, Specifications and Tolerances for Reference Standards and Field Standard Weights and Measures; 2. Specifications and Tolerances for Field Standard Measuring Flasks, National Institute of Standards and Technology Handbook 105-2, U.S. Government Printing Office, Washington, D.C., 1996. Available at http://www.nist.gov/pml/wmd

G. L. Harris, Specifications and Tolerances for Reference Standards and Field Standard Weights and Measures, 5. Specifications and Tolerances for Field Standard Stopwatches, National Institute of Standards and Technology Handbook 105-5, 1997. Available at http://www.nist.gov/pml/wmd

G. L. Harris, Specifications and Tolerances for Reference Standards and Field Standard Weights and Measures, 6. Specifications and Tolerances for Thermometers, National Institute of Standards and Technology Handbook 105-6, 1997. Available at http://www.nist.gov/pml/wmd

M. W. Jensen and R. W. Smith, The Examination of Weighing Equipment, National Institute of Standards and Technology Handbook 94, U.S. Government Printing Office, Washington, D.C., 1965.

G. D. Lee, Examination Procedure Outlines for Commercial Weighing and Measuring Devices, National Institute of Standards and Technology Handbook 112, 2002. Available at http://www.nist.gov/pml/wmd

Rand Corporation. A Million Random Digits with 100,000 Normal Deviates, Glencoe, IL: The Free Press, 1955. The Rand Corporation, 1700 Main Street, P.O. Box 2138, Santa Monica, California 90407-2138. Available at http://www.rand.org/publications/classics/randomdigits

Standard Method of Test for Density of Plastics by the Density Gradient Technique, ASTM D 1505-03, 2003. Available at http://www.astm.org

Standard Method of Test for Volume of Processed Peat Materials, ASTM D 2978-71, 1998. Available at http://www.astm.org

Standard Method of Test for Yarn Number by the Skein Method, ASTM D 1907-01, 2001. Available at http://www.astm.org

Standard Practice for Calibration of Laboratory Volumetric Apparatus, ASTM E 542-01, 2001. Available at http://www.astm.org

Standard Specification for Glass Volumetric (Transfer) Pipets, ASTM E 969-02, 2002. Available at http://www.astm.org

Standard Specification for Laboratory Glass Graduated Burets, ASTM E 287-02, 2002. Available at http://www.astm.org

Standard Specification for Polyethylene Film and Sheeting, ASTM D 2103-03, 2003. Available at http://www.astm.org

Standard Specification for Polyethylene Sheeting for Construction, Industrial, and Agricultural Applications, ASTM D 4397-02, 2002. Available at http://www.astm.org

Standard Test Methods for Thickness of Solid Electrical Insulation, ASTM D 374-99, 1999. Available at http://www.astm.org

U.S. Department of Defense Military Standard, Sampling Procedures and Tables for Inspection by Attributes (MIL-STD-105 D), U.S. Government Printing Office, Washington, D.C., 1963.

B. Younglove and N. Olien. NBS Technical Note 1079 – Tables of Industrial Gas Container Contents and Density for Oxygen, Argon, Nitrogen, Helium, and Hydrogen, 1985. Available at http://www.nist.gov/pml/wmd